Pyramid Power II

The Scientific Evidence

G. Patrick Flanagan, Ph.D.

PHI SCIENCES PRESS
Cottonwood, AZ 86326

Copyright © 2017 by Joseph A. Marcello, Ed. & G. Patrick Flanagan Ph.D. All rights reserved, foreign and domestic. No part of this book may be reproduced in any form, or by any means, mechanical or electronic, without permission of the copyright holder.

ISBN-13: 978-1542682176
ISBN-10: 1542682177

Printed in the United States of America

Note to Reader:

The manuscript that follows is an exact facsimile edition reproduced directly from a copy of Dr. Patrick Flanagan's original 1981 edition of *Pyramid Power II*. The task of recreating the many graphs, charts and tables of the original manuscript would have made the cost of re-setting the book prohibitive, as well as further courted the possibility of textual errors creeping in where none previously existed; hence, there may be occasional variations in clarity and symmetry. The only changes reflected are those with reference to new publishing and copyright information.

Joseph A. Marcello
Editor, *The Flanagan Revelations*
January 22, 2017

PYRAMID POWER II
SCIENTIFIC EVIDENCE

EDITED BY:
G. PATRICK FLANAGAN, PH.D.

PUBLISHED BY:
INNERGY™ PUBLICATIONS
P.O. BOX 18224
TUCSON, AZ 85731

In Collaboration with:
Mankind Research Foundation
1110 Fidler Lane, Suite 1215
Silver Spring, MD 20910

ACKNOWLEDGMENTS

We would like to gratefully acknowledge the participation of the following persons in this research project:

    Dr Theodore Horner
    Dr Joel Arem
    Dr Sandor Holly
    Dr Colin Frank
    Dr Michael Fineberg
    Dr Robert Edson
    Mr Leland Bristol
    Ms Skaidrite Fallah

In addition, special appreciation is extended to Boris Vern, Engineer Karel Drbal, Mr Elmer D. Robinson, Dr John Stauch, and Ms Carolyn Amundson for their helpful assistance and guidance on this effort and associated tasks in pyramid research.

DEDICATION

This book is dedicated to:

Ms Frances von Sacher

PYRAMID POWER II - PREFACE

PYRAMID POWER

If we consider the concept of Pyramid Power from a logical, rational frame of reference, we must look for an energy source which could explain some of the observed effects. This energy source must originate from within or without the structure, or both. In any case, an energy source must be located to explain the observed effects.

In recent years, Dr F.A. Popp of the University of Marburg in Germany has discovered that all living cells emit coherent laser light energy in the visible spectrum. These bursts of light are of high intensity, but of low average frequency. The result is that these photons of light cannot be detected without the aid of extremely sophisticated equipment. He has found that these emissions function in the living system in three distinct ways, (1) for the photo repair of DNA; (2) for photo repair of polypeptides, and (3) for intra cellular communications. His research further shows that living cells respond to and store photon energy from external sources, and that the incoherent energy is not recognized by the living system (see Electromagnetic Bio-Information, F.A. Popp). Dr Popp has further postulated that the living system probably emits and responds to coherent radiation across the entire electromagnetic spectrum, from ELF (Extremely Low Frequencies) to the Cosmic Ray band.

WHAT IS PYRAMID POWER?

The concept of Pyramid Power is based on the observation that foods and other items placed in a small model of the Great Pyramid of Egypt are affected in a positive way. Food items dehydrate without spoiling, the growth rate of plants is increased, the growth rates of bacteria cultures are decreased, the growth of crystals from supersaturated solutions is increased, the flavor of wine and other beverages is improved, and people sleeping in pyramids report heightened states of awareness, increased energy, and a heightened sexual response.

Since the publication of the book **Pyramid Power** by Dr Patrick Flanagan in 1973, the subject of pyramid power has been a raging controversy in the press and scientific circles.

Pyramid Power II is based on the first scientific study of the pyramid effect performed by independent scientific consultants. The research experiments presented in this volume were prepared under exacting scientific protocols, and executed by highly respected scientists who were hired as consultants by **Mankind Research Foundation**, in Washington, D.C. MRU is a non-profit research foundation directed by Dr Carl Schleicher.

We hope the experimental data presented in Pyramid Power II will stimulate others to perform exacting research into the pyramid effect.

TABLE OF CONTENTS

```
------------------------------------------PAGE
```
Preface................................... 1
Abstract.................................. 6
List of Tables............................ 9
I Introduction............................10
II Research Approach and Procedures......14
III Discussion of Results................28
IV Sleep Experiment Evaluation...........50
V Conclusions............................56
APPENDICES

Data Processing Plan.....................59
Experimental Data........................67
Examples of Statistical Calculations.....84
Experiment 13: Meditation in Pyramids...89
Subjective Observations..................95

Notes from Karel Drbal...................97
   Partial Patent Translation –
   pyramid razor blade sharpener.........99

   In depth Description of Pyramid
   Razor Blade Sharpening Theory........106

   Method of Calculating Electro-
   Magnetic Resonance of Pyramid
   Model................................123

Elmer D. Robinson:
   The Great Pyramid
   Its Design Concept...................124

   Geometry of the Great Pyramid........148

Biographies.............................156

LIST OF PHOTOGRAPHS

PYRAMIDS AND CUBES USED IN EXPERIMENTS
Between Pages 16 and 17

PYRAMID TENT
USED FOR SLEEP AND MEDITATION
Between Pages 92 and 93

MEDICAL EQUIPMENT
USED TO EVALUATE BIOLOGICAL PARAMETERS
Between Pages 92 and 93

BLOOD PRESSURE TEST
AFTER PYRAMID TENT MEDITATION
Between Pages 92 and 93

PREFACE

PYRAMID POWER II - PREFACE

PYRAMID POWER

If we consider the concept of <u>Pyramid Power</u> from a logical, rational frame of reference, we must look for an energy source which could explain some of the observed effects. This energy source must originate from within or without the structure, or both. In any case, an energy source must be located to explain the observed effects.

In recent years, Dr F.A. Popp of the University of Marburg in Germany has discovered that all living cells emit coherent laser light energy in the visible spectrum. These bursts of light are of high intensity, but of low average frequency. The result is that these photons of light cannot be detected without the aid of extremely sophisticated equipment. He has found that these emissions function in the living system in three distinct ways, (1) for the photo repair of DNA; (2) for photo repair of polypeptides, and (3) for intra cellular communications. His research further shows that living cells respond to and store photon energy from external sources, and that the incoherent energy is not recognized by the living system (see <u>Electromagnetic Bio-Information</u>, F.A. Popp). Dr Popp has further postulated that the living system probably emits and responds to coherent radiation across the entire electromagnetic spectrum, from ELF (Extremely Low Frequencies) to the Cosmic Ray band.

PYRAMID POWER II - PREFACE

Dr Phillip S. Callahan at the University of Florida has discovered that insects communicate to each other by means of coherent maser and laser radiation in the far infra-red band. His research has led him to believe that other living creatures do likewise.

One of the theories for explaining the pyramid effect involves the possibility that the pyramid structure acts as a reflector, and as such will refocus energy fields emitted by items placed inside the structure back into the items so that the energy could be re-used. This theory is clearly explained in the abstract of this paper. The recent discoveries by Popp and Callahan tend to support this thesis.

The other pyramid energy theory involves the action of the pyramid as a 'tuned' waveguide or cavity resonator. In this theory, random energy fluctuations of cosmic origin (outside of the earth) are assembled by the pyramid structure into a coherent standing wave of energy whose wavelength is determined by the geometry and physical dimensions of the structure. This externally concentrated energy field would then have an effect on substances treated inside.

The Czechoslavakian Engineer Karel Drbal has indeed demonstrated that the pyramid shape acts as a high efficiency electromagnetic resonator. He further

PYRAMID POWER II - PREFACE

proved that the shape could focus enough energy to create powerful physical effects on both organic substances and inorganic crystals such as that found in high quality razor blades.

Our own research at the Innergy™ Laboratories in Tucson, Arizona has shown that the pyramid structure acts as a high efficiency cavity resonator for both acoustic and electromagnetic energies. Our most recent research indicates that pyramids of the same geometry but different sizes create different effects.

Historically, it takes 50 or more years for a new discovery or idea to become accepted by established leaders of science. For example, the basic formula governing the basic law of electricity was discovered by Professor Georg Simon Ohm in the early 1800's. The law was so simple that a 12 year old child could perform the necessary experiments to demonstrate its accuracy. The leading scientific societies of his day categorically rejected his papers on the subject, claiming that his discovery was in the occult and was therefore scientific heresy.

Professor Ohm decided to privately publish a small booklet which demonstrated his law. When the scientific community learned of this they caused poor Ohm to lose his Professorship and made sure he could not work anywhere in Europe. He was eventually thrown into debtor's prison

PYRAMID POWER II - PREFACE

where he committed suicide.

Today, OHM's LAW is basic to ALL modern electronic technology. Without it we would still be living in the dark ages. The story of Ohm's Law is just one of dozens of presently accepted laws that were once relegated to the world of the occult.

When we published the book Pyramid Power in early 1973, a new scientific controversy came into being. Since that time, nearly 100 books have been published on the subject, and hundreds of pro and con news articles have appeared. It is interesting to note that a few prominent scientists including Carl Sagan have flatly stated that pyramids could not possibly produce any effects on items placed inside. An objective scientist would simply state that he had no opinion of the subject as he has not adequately researched it. Unfortunately, we all tend to fall into the trap of making instant judgments about things of which we have little or no knowledge.

There are dozens of claims for pyramid effects that are not supported by any type of valid research. Unfortunately it is hard to separate fact from fiction when it comes to pyramid claims. Most pyramid researchers are not trained in the scientific method, and therefore do not know how to set up proper research protocols.

PYRAMID POWER II - PREFACE

In the mid-1970's, Mankind Research Foundation, with the financial aid of Ms Frances von Sacher, hired a dozen independent scientists and testing laboratories to conduct a controlled study on the alledged pyramid effect. The scientists chosen for the job were among the best in their respective fields. These men have impeccable credentials, and have all published in numerous journals of science.

The research protocols for the various projects were carefully designed to ensure that the testing was done according to well established scientific methodology.

We hope this publication will stimulate others to conduct similar research in this area, and that the research protocol used in these experiments will serve as a guide to others who are not trained in the scientific method.

            G. Patrick Flanagan, Ph.D.
            Tucson, Arizona
            February, 1981

PYRAMID POWER II - ABSTRACT

ABSTRACT

Science is the attempt of intellect to set into order the facts of experience. The experience of many men, from the ancient Egyptians to modern investigators, have shown that pyramid structures may have an effect on the biological, physical, and psychological processes that are constrained to operate within the structure.

This research project documents effects, if any, on selected biological, physical, and psychological systems resulting from the placement of the system within a special rectangular pyramid shaped-structure. The faces of the structure are constructed of reflective materials, e.g., glass mirrors and aluminum foil bonded to a mylar backing; with the orientation of the reflective surface towards the center of the structure.

The pyramid face materials were selected so as to reflect any form of energy that may be given off by the system back to the system, hence minimizing such energy loss to the environment. Since the flow of energy from a local system to the environment is a measure of the <u>entropy</u> of the system, the entropic flow rate of those systems placed within the pyramids will decrease. It is then the effect of the decrease in normal entropic flow rate which

PYRAMID POWER II - ABSTRACT

is to be considered. Life processes are in general anti-entropic or negentropic, in that as growth proceeds, the entropy of the system decreases. The effect of the reflective pyramid structure should, from a theoretical standpoint, have a positive effect on life, increasing the growth rate of the living system. For dead organic matter undergoing entropic decay, the rate of decay should be decreased. Physical systems, on the other hand, are entropic and the effect of the pyramid structure should result in a decrease of the reaction rate, that is, a decrease of the reaction component in a direction corresponding to entropic flow. Psychological systems of the mind are the premier anti-entropic processes in the universe, so that the effect of pyramid structures are anticipated to increase the efficiency of the mental processes.

These experiments have been conducted to document the investigation of the hypothesis that placing a system within a reflective pyramid structure will have <u>an effect on that system</u> as compared to (1) a cubical structure of similar material and equal volume, and (2) an uncovered or zero structure situation. Emphasis has been placed on incorporating strict laboratory proceedure and statistical evaluation of data.

The results of the investigation are as follows:

PYRAMID POWER II - ABSTRACT

(1) The growth rates of black-eyed peas and lima bean sprouts are <u>increased</u>.

(2) The decay rates of dead organic systems such as raw meat, bananas, and eggs have been <u>decreased</u>.

(3) The growth rates of bacteria in cultures are <u>decreased</u>.

(4) The decay rates of plants isolated from light and water addition are <u>decreased</u>.

(5) The growth of crystals from supersaturated solutions is <u>increased</u>.

(6) The settling rates of colloidal suspensions are not affected within the measurement constraints.

(7) The subjective results of sleeping within a pyramid appear to indicate that less sleep is needed, dream states are heightened, and the general level of body energization is increased.

LIST OF TABLES

		PAGE
1	Description of Pyramids and Control Cubes used in each experiment	17
2	Description of collected data	26
3	Summary of results for experiments 01 through 12	31
4	Summary of results for Sleep Experiment	37
5	Sleep Experiment Day by Day Variations	41
6	Graph of Day by Day Variations	43
A1	Card Formats for Experiments 01 through 07	65
A2	Card Format for Experiment 08	66
A3	Summary of results for Meditation Experiment	93

PYRAMID POWER II - INTRODUCTION

I. INTRODUCTION

The belief that a pyramid structure appears to have special biophysical or phenomenal properties has come down to us from earliest historic times. (3000 B.C.) through such mediums as legends, myths, and religious documents including the Bible and the <u>Papyrus of Ani</u> (Egyptian Book of the Dead). There is clear evidence that the pyramids have been used as tombs, temples, and even energy sources, and it is this third aspect that is of primary interest here.

In 1900, Antoine Bovis, a French scientist, is reported to have found mummified animals near the King's Chamber of the Great Pyramid of Cheops at Giza, Egypt. He performed subsequent experiments with a three foot replica in the same proportions as the Cheops pyramid and found similar mummification (dehydration without putrefaction) using dead cats, raw meat, and raw eggs. In 1936, Dr Paul Brunton spent a night in the King's Chamber of Cheops and reported a mystical cosmic consciousness experience described in <u>A Search in Secret Egypt</u>. A Czechoslavakian engineer, Karel Drbal, duplicated Bovis's experiments and eventually patented a pyramid device in 1959 to sharpen (regenerate) razor blades.

Luis Alvarez, a Nobel Prize winning biochemist, measured the bombardment of the

PYRAMID POWER II - INTRODUCTION

Chefrens Pyramid (located in close proximity to Cheops) with cosmic rays in order to locate hidden chambers and found varying results when he tried to reduce the data, which prompted him to state that "something outside of known science is happening". Dr Patrick Flanagan in the late 1960's and early 1970's performed experiments in models of the Cheops pyramid and found that pyramids are a very powerful source of 'biocosmic' energy and that organic material does in fact dehydrate without putrefaction when kept inside a pyramid.

Other researchers as well, including Dr Boris Vern of Mankind Research Foundation (MRF), Washington, D.C., and Frances von Sacher of Connecticut, have reported similar effects of placing materials within scale models of the Great Pyramid of Cheops.

There is a causal relationship between the size, shape, and construction material of a structural enclosure and the chemical, physical, and biological processes which are constrained to operate within the enclosure. In this experiment the focus is on pyramid and cube-shaped structures.

There appear to be two separate but related effects which contribute to the total 'effect' of structure configurational energy relationships. There is the property of the structure to act as a wave guide to

PYRAMID POWER II - INTRODUCTION

capture, focus, and transduce cosmic energy of the electromagnetic spectrum (energy originating from outside the earth system). A structure such as a pyramid chamber acts as a resonator to collect energy and focus this energy. An example of a structure which 'captures' energy is the crystalline antenna of a crystal radio. This crystal captures the radio frequency electromagnetic energy, amplifies it, and transduces it to electrical signals which power the speaker. The second effect is the use of a shape which has highly reflective inside material so that almost all energy generated inside is kept inside and is not lost to the environment. Energy is thus added to that of the internal system by recycling of the energy given off by the system.

In both cases one sees an <u>increase in energy</u> for any system which is placed within the pyramid structure. The question then becomes, does the existence of the added energy reinforce or cancel the energy processes naturally occuring within the pyramid structure? Effects of this added energy influx to various systems by use of reflective pyramid structures has been documented in hopes of answering this question and hopefully to note the general characteristics of systems whose energy processes are positively reinforced and those which are cancelled. Note that in general the flow of energy through a system tends to organize the system. This may be interpreted as saying that in general the

PYRAMID POWER II - INTRODUCTION

negentropy (order tendency) of a system is increased or that entropy (disorder tendency) of a system is decreased. The question then is rephrased in these entropic terms.

The experiments fall into three categories: (1) biological systems and processes, (2) physical and chemical processes, (3) psychological systems. Experiments were conducted in each area as described in the following table:

BIOLOGICAL

Meat Decay and Dehydration
Bacteria Growth
Banana Decay and Dehydration
Root Growth

PHYSICAL

Crystal Growth
Colloidal Suspension Settling

PSYCHOLOGICAL

EFFECTS OF SLEEP STATE

II. RESEARCH APPROACH
AND
EXPERIMENTAL PROCEDURES

The basic form of pyramid research experiments was to place one object within the pyramid (located at the center of the volume of the pyramid) and to place two control objects, one covered with a cubic structure of equal volume to the pyramid, and one uncovered object outside the pyramid in close proximity. The objects, which are initially identical, will be monitored at certain specified time intervals to ascertain through careful laboratory measurement the differences in the states of the objects, if any.

The initial set of experiments will determine if there is any measureable effect due to the special pyramids employed with respect to cubic enclosures and to a lack of enclosure. It must be determined whether the effects associated with pyramid shapes are simple enclosure effects.

The rectangular pyramids used in these experiments are glass structures which are lined with a reflective surface oriented inwards. In effect, the pyramids are constructed of mirrors. There are three sizes of experimental pyramids with base sides of 6 inches, 10 inches, and 16 inches, and heights of 3.81 inches, 6.36 inches, and 10.08 inches respectively. A large pyramid of mylar backing covered with

PYRAMID POWER II - RESEARCH APPROACH

a bonded layer of aluminum with dimensions of six feet along the base and 3.81 feet high is used for the sleep and meditation experiments.

The cubic control structures are built of glass and in many cases are lined with a reflective inner layer oriented so that the reflection is inward. The volume for each experimental cube is equal to that of its associated pyramid.

The data is evaluated statistically, where applicable, to determine if there is in fact a difference between the treatments and which will give a confidence level for stating that a difference exists.

A. EXPERIMENTAL DESIGN

The units for the experimental design of the pyramid investigations consisted of plants, human test subjects, pieces of meat, and other biological entities. An experimental unit is defined as that physical entity on which an action (treatment) of some kind is imposed. In the present instance, there were three treatments:

1. PYRAMID

Storage of the experimental unit within a pyramid-shaped structure.

2. **CUBE**

Storage of the experimental unit within a cube of equal volume.

3. **UNCOVERED**

Uncovered exposure of the experimental unit.

PYRAMIDS AND CUBES

The photograph above illustrates examples of the pyramids and cubes which were used in the various research experiments.

TABLE 1. DESCRIPTION OF PYRAMIDS AND CONTROL CUBES USED IN EACH EXPERIMENT

		Size of Pyramid		Cube
Experiment	Name	Base Length	Height	Base Length
01	Black-eyed Peas	10"	6.36"	6.8"
02	Lima Beans	10"	6.36"	6.8"
03	Hamburger Meat	6"	3.81"	4.1"
04	Throat Bacteria	6"	3.81"	4.1"
05	Rooting	10"	6.36"	6.8"
06	Crystal	6"	3.81"	4.1"

TABLE 1. DESCRIPTION OF PYRAMIDS AND CONTROL CUBES
CONTINUED

Experiment	Name	Size of Pyramid		Cube
		Base Length	Height	Base Length
07	Yeast Bacteria	6"	3.81"	4.1"
08	Sleep	6'	3.81'	4.1'
09	Live Plant	16"	10.08"	11"
10	Bacteria on Egg	10"	6.36"	6.8"
11	Banana	16"	10.08"	11"
12	Colloidal Suspension	10"	6.36"	6.8"

PYRAMID POWER II - RESEARCH APPROACH

A randomized complete block design was employed for all treatments with three experimental units per block, the treatments being assigned at random to the three experimental units. In addition to the foregoing randomization, the location of the experimental units on the bench and other relevant randomizations were performed as required. Randomization insured that the experimental units with the several treatments were randomly interspersed throughout the experimental space so that the external factors had an equal chance of influencing the several treatment effects.

The null hypothesis investigated for the several experiments indicated that the three treatments had similar effects. In the testing of this hypothesis, the blocks were assumed to be random. Standard statistical methodology was employed in the testing of the null hypothesis. Example computations for several of the experiments are given in the appendices. The analyses of the remaining experiments were accomplished by computer.

B. EXPERIMENTAL DESCRIPTION

The first round of experiments described in this section were performed at the Physicians Bioanalytical Laboratory, Suitland, Maryland during January of 1975. This is a licensed laboratory which is fully equipped to carry out the experiments specified in the research protocol.

PYRAMID POWER II - RESEARCH APPROACH

However, due to the inability to maintain strict controls over temperature and other environmental factors at the Physicians Analytical Laboratory, the second and subsequent experimental rounds were shifted to the facilities of the Mankind Research Foundation (MRF), Washington, D.C. MRF is located in a modern building with automatic environmental controls for temperature, ventilation, and humidity. Experiments were conducted during the period of February - April 1975. Since only the bean sprout experiments were performed during the first round, which are relatively immune to minor environmental changes, it is not considered likely that the experiments conducted are affected by external factors to the extent of introducing bias into the experimental design. The data collected did not appear to reflect such bias and all data upon which statistical analysis was performed was considered valid.

This section contains experimental procedures, followed by a table of measurements for each experiment. Note, however, that certain experiments do not allow for statistical analysis.

01 and 02. <u>Growth of Bean Sprouts</u>. In each treatment a number of beans were placed upon a sponge which had been soaked with a prescribed amount of water (50 ml) and covered with a wet paper towel. The sponges were assigned randomly to locations and treatments were also assigned randomly

PYRAMID POWER II - RESEARCH APPROACH

to locations. The pyramid and cubic structures were sealed from the atmosphere with duct tape. After five days, the length of the bean sprout root was measured from a reference point defined as the point at which the root left the bean.

03. <u>Decay of Hamburger</u>. Ten grams of fresh hamburger meat was placed in each of 12 Petri dishes and the dishes were divided into four blocks of three dishes per block. Treatments were assigned randomly in each block. The pyramid and cube were sealed with duct tape to the table so that no air could enter or escape. After five days, observations were made and subjective ratings of each dish were made. The dishes were observed by various random individuals and were shuffled on the table so no one knew which dish they were evaluating as far as treatment was concerned. The result was a rating on a scale of 1 to 5, with 1 representing perfectly preserved meat and 5 representing decayed, hard, spoiled meat; hence data for each group consists of a number between 1 and 5. This group of four blocks was repeated, giving a total of 8 blocks.

04. <u>Bacteriological Growth</u>. A throat culture was taken and placed in a test tube with a liquid culture material and allowed to grow for two days in a 90 degree F heated room. This resulted in the growth of bacteria randomly distributed throughout the liquid. Six blocks of culture material conisting of glucose

PYRAMID POWER II - RESEARCH APPROACH

(agar) and sheep's blood were inoculated. A special inoculation tool composed of a wire rod with a small loop at the end was dipped into the liquid culture and then placed on the culture material in a repeated pattern. The pattern was a square of small tangent circles formed by placing five circles on a side, making a pattern of 25 small circles. One more loopful of bacteria was placed within this pattern and spread around as equally as possible. The plates were randomly assigned as to location, and was treated to location within the block of three. The number and size of the growths were measured.

05. <u>Plant Rootings</u>. A Wandering Jew plant was divided into six cuttings by severing stems of the many-stemmed plant about an inch below the the lowest of the leaf structures. The bottom leaf was removed, leaving an exposed cylindrical area around the stem through which the main roots would appear. The cuttings were placed in small jars of water, making sure the band was covered by an inch or more of water. The pyramid and cubic structures were sealed from the atmosphere. The number and length of each root was measured after six days.

06. <u>Crystal Growth</u>. Supersaturated solutions were made by heating water to the boiling point and dissolving as much of the solid chemical potassium aluminum sulfate (approximately 16 oz.) as possible. For block 1 and 2, 20 ml of this supersaturated

solution was placed in Petri dishes, assigned to location and treatment and allowed to cool for 8 hours. As this solution cools, crystals are spontaneously nucleated. The weight of crystals formed and the shape of the crystal were determined.

07. <u>Yeast Bacteria</u>. A mixture of Brewer's Yeast (1/4 oz.) was mixed with one half cup of 100 degree F water and allowed to grow for six hours. Eighteen culture dishes of sheep's blood and agar were innoculated with approximately equal amounts of the yeast solution and allowed to grow for 24 hours in a 90 degree F room. The 18 dishes were then divided into six blocks of three each in which the area of the yeast growth was equal. The three dishes within each block were randomly assigned to location and treatment and the size of certain specific cultures within the yeast growth in each dish was measured. The dishes were treated for six days and the size of the certain specific cultures was measured again. The area of 'specific' culture for each dish was calculated before and after treatments were applied and a ratio of the post to pre-treatment areas was calculated. These area ratios were then normalized with respect to the mean area ratio of the uncovered treatment.

08. <u>Sleep State</u>. Subjects slept in one of two treatments -- an uncovered, normal situation and in square pyramid tents measuring six feet on a side and 3.81

PYRAMID POWER II - RESEARCH APPROACH

feet high. The six day sleep experimental period was divided into two parts -- a two day period in which the subjects slept in the untreated control situation and a four day period in which the subjects slept in the pyramid tent. Measurements of Blood Pressure, Galvanic Skin Response, Pulse Rate, and Body Temperature were taken before sleep at night and in the morning upon waking.

09. <u>Live Plants</u>. Six healthy live Nephthytis plants were divided into two blocks of three plants each and treatments within blocks and locations within blocks were randomly assigned. The pyramid and cubic structures were sealed from the atmosphere with duct tape. Subjective analyses of the plants were performed after a period of 21 days. There was no addition of water or nutrient food, and neither the pyramid nor cubic treatments allowed light to enter.

10. <u>Eggs</u>. Four eggs were cracked open, placed in uncovered Petri dishes, treated with equal amounts of solution containing highly concentrated bacteria, assigned randomly to four treatments, and sealed from the atmosphere. The treatments included the three previously noted and a fourth consisting of a plexiglas pyramid of equal volume to the mirror pyramid and the cube. The bacteria growth in terms of number and size of colonies was measured after five days.

PYRAMID POWER II - RESEARCH APPROACH

11. <u>Bananas</u>. Six bananas were divided into two blocks of three treatments, randomly assigned, sealed from the atmosphere, and left for ten days. Subjective analyses were performed.

12. <u>Colloidal Suspension</u>. 1.5 grams of solid potassium ferrocyanide and 1.5 grams of solid ferric nitrate were each dissolved in 300 ml of water. For block 1, 100 ml of potassium ferrocyanide and 50 ml of ferric nitrate were thoroughly mixed and equal volumes (50 ml) were placed in three glass beakers. The beakers were randomly assigned to location and treatment and the colloidal precipitate of the chemical mixture was allowed to settle. The amount of settling as measured by the color of the remaining unsettled colloidal solution was noted after one, three, and six hours.

Block 2 repeated the experiment with 300 ml of potassium ferrocyanide and 15 ml of ferric nitrate being mixed, divided equally into three beakers (105 ml each), randomly assigned to location and treatment, and allowed to settle. Measurement was made after one and three hours. In Block 3, 300 ml of potassium ferrocyanide and 3 ml of ferric nitrate were mixed, divided, assigned, and allowed to settle. Measurements were taken after one and three hours.

Table 2 summarizes the characteristics and form of the data collected for each experiment.

PYRAMID POWER II - RESEARCH APPROACH

TABLE 2

DESCRIPTION OF COLLECTED DATA

Experiment			Form of Data	
Number	Name	Characteristic	Structure	Type
01	Black - eyed peas	Length of Sprout	x.xx	Number of inches
02	Lima beans	Length of Sprout	x.xx	Number of inches
03	Hamburger	Freshness	x	5-Point Scale
04	Throat Bacteria	Area	xxx	Normalized Area
05	Rootings	Sum of Root Length	xx.x	Length in inches
06	Crystal	Weight of Crystal Formed	xx.x	Measured in gms
07	Yeast Bacteria	Area	x.xx	Normalized area

26

TABLE 2
CONTINUED

Experiment Number	Name	Characteristic	Form of Data Structure	Type
08	Sleep	Blood Pressure		
		Systolic	xxx	mm of Hg
		Diastolic	xxx	mm of Hg
		Pulse rate	xx	Cycles/Minute
		GSR	x.x	Normalized No.
		Body Temperature	xx.x	Degrees F
		(The sleep measurements were taken both before and after treatment)		
09	Live Plant	Subjective analysis of health and vitality		
10	Bacteria on Egg	Area of Bacteria Growth	x.xx	Normalized No.
11	Banana	Subjective analysis of preservation		
12	Colloidal Suspension	Time of Settling	xx	Seconds

PYRAMID POWER II - RESULTS

III. DISCUSSION OF RESULTS

The treatment mean deviations for the pyramid experiments are given in Table III along with the probability that the observed differences between the means could have arisen by chance alone.

For example, Experiment 01 relating to the length of sprouts of black-eyed peas, showed means of 1.52 and 1.01 for the pyramid and cube treatments respectively, there being no uncovered treatment. The observed difference could have arisen by chance alone with a probability of 0.129, or about 1 chance out of 8. Since both the pyramid and the cube structures were of similar materials, opaque to light, and sealed to the table top so as to retain moisture, it would appear that the difference in treatment means (assuming that a 1 in 8 probability is considered unusual) is in fact due to the differences between the two treatments and not due to differences in moisture absorption or loss, or in the amount of light present.

Experiment 02, the growth of lima bean sprouts, shows the same qualitative results as Experiment 01 in that the length of the sprout is, on the average, 1.64 inches for the pyramid treatment and 0.68 inches for the cubic control. However, due to the limited number of replication blocks the probability is 0.284 or approximately 1

PYRAMID POWER II - RESULTS

chance in 4 of achieving the observed results by chance. This again indicates (with the assumption that a 1 in 3 chance is considered unusual) that sprout growth is enhanced by a pyramid structural enclosure.

The results of Experiment 03, the preservation of raw hamburger meat, are most impressive in terms of the statistical significance. It is shown that the meat is relatively preserved, as measured on a freshness scale of 1 to 5, with respect to a cubic enclosure and uncovered situation at a significance level of 0.001. The probability of the observed results occurring as a matter of chance is less than 1 in 1,000. On a scale of 1 to 5, 1 represents perfect preservation and 5 represents dehydration and putrefaction. On the basis of this scale, the pyramid treatment showed a mean value of 2.28, the cubic treatment indicated a value of 4.62, and the uncovered treatment a value of 4.64. The result is consistent with Experiment 04 and 07, which showed that the growth rate of bacteria was decreased by pyramid treatment. Since decay in meat is partly a bacterial growth process, this bacterial growth in decaying meat is decreased, hence a relative state of preservation is achieved.

Experiment 04 is the effect of the treatment on the growth rate of throat bacteria. The results indicate at a significance level of 0.05 that bacterial

PYRAMID POWER II - RESULTS

growth rate was decreased. In terms of percentages with respect to the uncovered treatment, the amount of growth in the pyramid treatment was 48 percent and the cubic treatment growth was shown to be 94 percent.

TABLE 3

SUMMARY OF RESULTS FOR EXPERIMENTS 01 THROUGH 12

Treatment Mean Deviations

Experiment Number	Name	No. of Blocks	Pyramid	Cube	Uncovered	Probability
01	Black-eyed peas Sprout length "	6	1.90	1.27	—	0.06
02	Lima Beans Sprout length "	2	1.641	0.679	—	0.284
03	Hamburger Meat Freshness (0 – 5)	8	2.230	4.618	4.637	0.001
04	Throat Bacteria (Normalized Area with respect to area of uncovered treatment)	5	48%	94%	100%	0.001
05	Rootings (Volume of roots in 0.0001 cubic in.)	2	0.6	33.5	32.8	0.10
06	Crystal (Wt. in grams of crystal formed)	3	20.1	19.9	18.2	0.10<p<0.25

TABLE 3

CONTINUED

Treatment Mean Deviations

Experiment Number	Name	No. of Blocks	Pyramid	Cube	Uncovered	Probability
07	Yeast bacteria (Normalized area with respect to uncovered treatment)	6	50%	86%	100%	0.10
08	Sleep Ratio of AM to PM Measurements			Inside Tent	Outside Tent	
	Pulse			0.944	0.880	0.160
	Temperature			0.994	0.993	over 0.500
	GSR			1.263	1.681	0.174
	Systolic			0.966	0.926	0.092
	Diastolic			1.041	0.961	0.025
09	Live Plant	2	alive	dead	dead	———

32

TABLE 3

CONTINUED

Treatment Mean Deviations

Experiment Number	Name	No. of Blocks	Pyramid	Cube	Uncovered	Probability
10	Bacteria on Egg (Area of growth based on area of pyramid treatment as 1)	1	1,16*	100	none	———
11	Banana	1	fresh	decayed	decayed	
12	Colloidal Suspension	4	no noticeable difference			———

*Two pyramids were used, one mirrored, one plain. The mirrored bacteria area was taken as 1.

33

PYRAMID POWER II - RESULTS

Experiment 05 is the rooting of the Wandering Jew Plant. This experiment differs from the others in that the control cubic structure was not opaque to light and did not have a mirrored surface. The results, which are significant to a 0.10 level, tend to indicate that the pyramid treatment effectively blocked the rooting process from occurring. However, this may be due to the pyramid treatment structure blocking the plants from light sources, since light is obviously necessary for the rooting process. This area may need further investigation; since the pyramid treatment prevented the rooting process, then a state of decreased metabolic processes may be in evidence, although there was no evidence of leaf deterioration and plant decay or expiration. The mean value of root (based upon an estimation of the length and diameter of each rootlet) was calculated as follows:

Pyramid Treatment = 0.6×10^{-4} inches^3

Cubic Control = 33.4×10^{-4} inches^3

Uncovered Treatment = 33.8×10^{-4} inches^3

Experiment 06 evaluates the growth of potassium aluminum sulfate crystals in a

PYRAMID POWER II - RESULTS

supersaturated solution. The mean values of weight of crystals formed for each treatment are 20.1 gm for the pyramid, and 19.9 gm for the cubic treatment, and 18.2 gm for the uncovered treatment. The significance of these results in terms of the probability of pure chance is between 0.10 and 0.25, or between 1 in 10 and 1 in 4. This result can be interpreted as indicating that further experiments involving crystal growth would be helpful to verify these findings.

Experiment 07 is measurement of the growth of yeast bacteria. The results, which are significant to a probability of 0.10, show that the growth of bacteria is decreased. The mean value of treatments are 51 percent, 86 percent, and 100 percent for the pyramid, cubic, and uncovered treatments respectively.

Experiment 08 carried out the sleep experiment, and in general it was observed that all variables* showed the expected trends with respect to differences between AM and PM measurements. Pulse rate decreased, GSR readings increased, systolic blood pressure decreased, and diastolic blood pressure decreased. However the pyramid treatment differences were less than the differences recorded in regular sleep.

The following table represents the summary of results for the sleep experiment and it can be observed that the Before

PYRAMID POWER II - RESULTS

Sleep or PM measurements for both treatments are roughly equivalent. It is in the Post Sleep or AM measurements that significant differences appear.

TABLE 4
Experiment 08
Summary of Results for Sleep Experiment
(Number of Blocks = 6)

Variable	Computer Variable	Treatment Means		Probability
		Inside Tent	Outside Tent	
BEFORE SLEEP DATA				
Pulse (count)	CHAR1(1)	79.83	74.17	0.001***
Temperature (Deg F)	CHAR1(2)	98.18	98.18	0.436
GSR (Normalized No)	CHAR1(3)	1.996	2.083	over 0.500
Systolic (mm of Hg)	CHAR1(4)	126.25	126.67	over 0.500
Diastolic (mm of Hg)	CHAR1(5)	78.29	79.17	over 0.500
AFTER SLEEP DATA				
Pulse	CHAR2(1)	74.33	64.17	0.012**
Temperature	CHAR2(2)	97.57	97.53	over 0.500
GSR	CHAR2(3)	2.496	3.192	0.187
Systolic	CHAR2(4)	121.8	117.0	0.203
Diastolic	CHAR2(5)	80.7	75.3	0.076

Probability of achieving observed differences by chance alone.
* means significant at the 0.05 level.
** means significant at the 0.01 level.
*** means significant at the 0.001 level.

TABLE 4
Experiment 08
Summary of Results for Sleep Experiment
CONTINUED

Variable	Computer Variable	Treatment Means		Probability
		Inside Tent	Outside Tent	
AFTER MINUS BEFORE				
Pulse	DIF(1)	-5.500	-10.00	0.137
Temperature	DIF(2)	-0.608	-0.650 over	0.500
GSR	DIF(3)	0.500	1.108	0.332
Systolic	DIF(4)	-5.417	-9.667	0.125
Diastolic	DIF(5)	2.417	-3.833	0.038*
RATIO OF AFTER TO BEFORE				
Pulse	RAT(1)	0.944	0.880	0.160
Temperature	RAT(2)	0.994	0.993 over	0.500
GSR	RAT(3)	1.263	1.681	0.174
Systolic	RAT(4)	0.966	0.926	0.092
Diastolic	RAT(5)	1.041	0.961	0.025*

Probability of achieving observed differences by chance alone.

* means significant at the 0.05 level.
** means significant at the 0.01 level.
*** means significant at the 0.001 level.

PYRAMID POWER II - RESULTS

The measured parameters are classic in the sense that they are standard physiological variables indicating an individual's readiness to respond to various stimuli. Each of the measures is susceptible to modulation by both the internal state of the organism (drugs, diseases, etc.) and by external motivation such as physical and psychological stress. In this study the physiological parameters are used to determine the effects on human readiness and potential due to an external condition or treatment. It is indicated that the pyramid treatment increases the individual's preparedness, energization or action potential level upon awakening.

It was further observed that as the body awakens, there is a higher degree of readiness to respond to outside stimuli. This is shown in the following demonstration.

A standard graph of performance quality (measure of the productive responses to external stimuli) versus arousal level follows the distribution shown below.

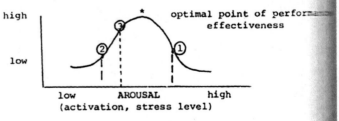

PYRAMID POWER II - RESULTS

      Based upon general observations, it can be noted from the graph that at the end of a day's activity, represented by the state at point (1) on the curve, regular sleep will result in a lowering of arousal and a release of built-up stress levels to the level of point (2). The observations regarding pyramid treatment sleep show that the level of arousal does not drop as low as that following untreated sleep and hence may drop only to point (3). The curve also suggests that as a result of pyramid sleep, the subjects tend to approach a state of optimal functioning. This is, of course, a very general treatment of sleep, but there are indications of some potential benefits and favorable effects of sleeping in pyramidal structures. Further research is, however, necessary and closer control and attention must be given to the many complex factors and variables involved in sleep, waking, and arousal states. The psychological data in terms of subjective evaluations and tests showed a direct correlation with the physiological data in respect to the energization effects of sleeping within pyramid structures.

TABLE 5

SLEEP EXPERIMENT

DAY-BY-DAY VARIATIONS

	1			2			3		
	PM	AM	DIF	PM	AM	DIF	PM	AM	DIF
Pulse	75.3	61.3	-14	73	67	-6	83.2	79.5	-3.7
GSR	1.8	4.4	2.6	2.4	2.0	-0.4	1.9	2.7	0.8
Systolic Blood Pressure	128	116	-12	125.5	118	-7.5	125	122	-2.5
Diastolic Blood Pressure	80	75	-5	78.5	76	-2.5	74	80	6

TABLE 5

SLEEP EXPERIMENT

DAY-BY-DAY VARIATIONS

	4			5			6		
	PM	AM	DIF	PM	AM	DIF	PM	AM	DIF
Pulse	80	69.7	-10.3	85.3	73.6	-11.7	71	74.7	+3.7
GSR	2.0	2.3	0.3	2.0	2.25	0.25	2.1	2.7	0.6
Systolic Blood Pressure	130	120	-10	130	119	-11	118	122	+4
Diastolic Blood Pressure	84	78	-6	79.5	83	3.5	75	82	7

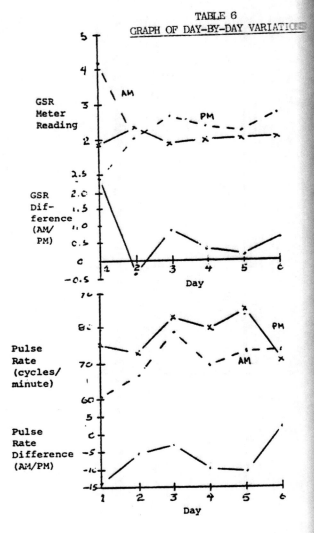

TABLE 6
GRAPH OF DAY-BY-DAY VARIATIONS

TABLE 6
GRAPH OF DAY-BY-DAY VARIATIONS

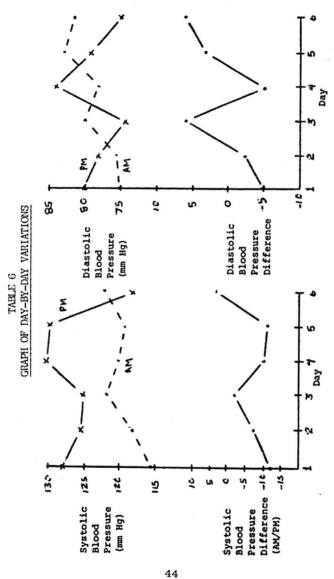

PYRAMID POWER II - RESULTS

A rise in pulse rate can indicate an enhancement of the body's potential for action or arousal level. During sleep this level drops considerably as arousal decreases; upon waking, the pulse level is generally below that necessary for efficient daily activities and hence the pulse rises. This period of pulse increase is a 'start-up' period. Since the pulse decreased due to pyramid treatment sleep, the amount (-5.5) is less than that observed during regular sleep (-10.0). The probability level of 0.137 shows that the pyramid sleep treatment may in effect be shortening the 'start-up' period experienced during the commencement of morning activities.

GSR or Galvanic Skin Response is a measure of the electrical resistance of the human body. Low values of GSR correspond to high stress or arousal levels. These measurements indicate that sleep increases the GSR reading, that is, the level of body arousal is decreased, but that pyramid treatment sleep does not increase the GSR (+0.50) as much as regular sleep (+1.11). With a probability level of 0.33, there is a 1 in 3 odds that these results are due to chance alone. Again, there is a suggestion that upon waking the arousal level or energization due to the pyramid treatment is greater than that of regular sleep in uncovered areas.

Similar effects were observed in the blood pressure measurements, with the

PYRAMID POWER II - RESULTS

exception that sleep tends to lower blood pressure. This was noted in the untreated sleep data as the systolic blood pressure dropped by 9.67 mm of Hg and the diastolic pressure dropped by 3.83 mm of Hg. The pyramid treatment sleep shows the expected decrease in systolic pressure (-5.42 mm Hg), but shows there is an increase in the diastolic pressure of +2.42 mm of Hg. The interpretation of this information is not discernable at this time, but it may represent some unknown type of energization process taking place. This decrease in blood pressure measurements was observed to be less for the pyramid treatment sleep than for the untreated sleep. The findings in terms of probability are significant in respect to the systolic results with a probability level of 0.125 as contrasted with the diastolic results at a level of 0.038.

Analysis of the pyramid treatment data suggests a cumulative effect which is compounded after several consecutive days of pyramid sleeping. The graphs which follow plot the observed data in a manner designed to show this trend.

Experiment 09 consisted of placing five Nephthytes plants (high water content) in three treatments and observing the condition of the plants after 21 days. There was no addition of water or nutrients, and neither the pyramid nor cubic treatments allowed light. The plants kept in the pyramid treatment were noted to

PYRAMID POWER II - RESULTS

be much healthier with respect to those in the cubic and the uncovered treatments.

OBSERVATIONS

Two plants under pyramids appeared healthy and green in the upper leaves; and stalks were upstanding, strong and healthy in appearance.

Bottom leaves were yellowed, but not wilted; appeared to be still moist and to have life, as if only some of the moisture and all of the chlorophyll in the bottom leaves were withdrawn into the top leaves. In general, both plants appeared to have grown in size under the pyramid, even though they were airtight. No moisture was condensed in the pyramids, but remained within the the plant. Soil in pots remained moist.

Two plants under cubes had wilted and dried; soil in pots was hard and dried out. Plants had almost died totally, most of the green color had faded in the upper leaves, and lower leaves were totally dead and yellowed.

It appears that the process

PYRAMID POWER II - RESULTS

incurred by plants as light is denied is slowed down by storing the plants in pyramid structures. The results may be expected if it is hypothesized that the life process is slowed down in some way. This may be due to the fact that the pyramid reflects energy back to the plant and it is probable that the plant is able to reuse this energy.

Experiment 10 is the growth of highly concentrated throat bacteria on raw egg yolks. The growth of the bacteria was much less in the mirrored pyramid and in the plexiglas control pyramid than in the cubic control structure. No bacteria grew on the uncovered egg as it quickly dehydrated. The amount of bacteria as measured by surface area of bacteria growth and normalized with respect to the area of the mirrored pyramid bacteria growth was 1.0 for the mirrored pyramid, 16 for the plexiglas and 100 for the cubic control structure.

Experiment 11: Banana Dehydration and Putrefaction

Subjective Banana Evaluations

Observer 1:

Banana under pyramid was firm and still edible, although the skin was brown. Banana smelled sweet.

PYRAMID POWER II - RESULTS

Banana under cube was rotten and softened, inedible.

Banana left out was totally rotten, even very gushy. Decomposed.

Observer 2:

Pyramid specimen -- Although skin was black and specimen had softened to some extent, it was obviously in a better state of preservation; not liquified or dark inside like cube specimen.

Experiment 12: Colloidal Suspensions

It was not possible to observe a difference in the settling rate of the various treatment colloidal suspensions.

PYRAMID POWER II - SLEEP EXPERIMENT

IV. SLEEP EXPERIMENT EVALUATION

In view of the special interest and significance of investigating the pyramid structures from the standpoint of sleep effects and altered states of consciousness, it is considered useful to include the senior psychologist's evaluation report. The results obtained during the sleep experiments should be viewed in terms of the comments and restraints indicated in this evaluator's report.

This evaluator's report describes psychological changes associated with sleeping underneath specially constructed pyramids. A group of seven people participated in an assessment of the effects of sleeping under small pyramids. In order to accomplish this task, the group was asked to sleep uncovered in the rooms where the pyramids would be set up for two nights. It was thought that these nights would serve as acclimatization to the unfamiliar situation. These two nights were followed by four nights under the pyramids. (One participant actually had a much longer experience under the pyramid, in excess of twenty days.)

The method of assessing psychological changes consisted of the following: There was a group interview before and after the six day on-site period

PYRAMID POWER II - SLEEP EXPERIMENT

of observation. In addition to daily physiological measures, each participant was asked to write about their experience during the previous 24 hours (or from the time of their last PM recording). The directions were to report any new experiences that had occurred, anything novel or unique. Each morning the participants were asked to report on their thoughts, feelings, and memories. Dreams were to be reported under the category of memories. In order to keep the analysis as bias-free as possible, each participant's written records were identified by a random number series to keep the judge from knowing their true sequence.

SUBJECTS. A profile of the group was developed from a questionaire and two interviews. The group was made up of two females and four males. One subject participant was excluded because of no data. They were all young adults with an average age in their late twenties. They were all bright and verbal and displayed no overt pathology. In most respects they were a normal or average group. However, they could not be considered a random sample. They all had some kind of expectation about the positive effects of the pyramids. This fact would suggest that there might be some bias in their self-report. At the same time, they all had an unusual amount of introspective ability, or were 'psychologically minded'. These factors obviously set them apart from a random sample of Americans. These factors could also be thought of as an advantage, since they all were relatively sophisticated in describing their own experience, and probably more sensitive to changes in their thoughts and emotions. One could infer that this group was less likely to defend against, or reject new experiential data out of hand.

PRE-/POST INTERVIEW DATA. As suggested above, all the subjects were ego-involved in the task. They were looking forward to the experience of sleeping in the pyramids. There were some reservations expressed about carrying out the project inside a building with a good deal of insulation from the external world, under controlled laboratory conditions. There was a concern for the energy that might be displaced coming from outside the pyramid, but little attention was given to the possibility of conserving, focusing, or otherwise intensifying internally generated energies. After the experience under the pyramid, there seemed to be a shift to an appreciation of the internally generated energies. More exacting interviews and testing would be needed to confirm this point, but it seems to be worth noting.

Another interesting pre/post change in the group was the intense involvement with the other participants. It was as though they had been through a marathon group therapy. It is difficult to disentangle cause from effect in this situation, but it would be worthwhile

PYRAMID POWER II - SLEEP EXPERIMENT

following up on the dimension of interpersonal closeness associated with the use of the pyramid.

The group was asked to identify dimensions that could be used in evaluating pyramid-sleep experiences. There were many suggestions, but a few seemed to qualify as consensus statements. First was the quality of sleep. Second was the vividness of dream recall. Third was overall emotional tone, from depression to elation. Fourth was openness to people and new experiences. There was the general opinion expressed that the pyramid experience produced change on these dimensions, but not necessarily in the same direction for all group members.

'DIARY' INFORMATION. A written record of experiences was requested from each of the six participants in the evening and morning of each day. There were six pairs of AM/PM records for each person. These were given a random number series code and given to the judge. His task was to identify if the records were from the two pre-pyramid AM/PM sequences, or from the four pyramid-use AM/PM sequences. This 'blind' analysis was carried out with the following results:

The AM/PM sequences for two participants were correctly identified, and another one was partially identified (one pre-pyramid AM/PM and three during pyramid AM/PM sequences). The other three individuals' records could not be correctly

PYRAMID POWER II - SLEEP EXPERIMENT

identified. Generally speaking, the judgements were made in terms of how 'good' the reported experiences were, reflecting more, perhaps, of the judge's bias and expectations than what was in the data.

After the random number codes were revealed, the AM/PM pairs were analyzed for common elements. These are the conclusions reached from several readings and reference to the dimensions derived from the group interviews. Vivid dreams and no dreaming were both reported. Some reports of no dreaming were accompanied by statements indicating that the person thought that they dreamt but did not remember the content. This suggests that with better observation techniques, it might be established that dreaming is enhanced.

There were statements about having been involved in emotional situations during the daytime hours. These were either being open to new experiences, or being involved as a third party in other people's emotional laden scenes, or actually being an actor in an emotion laden situation. Traditional scientific caution would forbid any lasting conclusions, but this one small sample does alert one to this possibility.

The one final point comes from a judgment of change on the part of the subjects. They seem to make statements that indicate a change of one's fixed patterns. The change seems to be unsettling of stable

PYRAMID POWER II - SLEEP EXPERIMENT

precepts. Again, one should not overgeneralize from such a small sample. The data are impressionistic and unsystematic and represent a single interpretation.

One point should be stressed equally with the conservative view of the experimental method: The participants agreed that there was some effect on them from having slept under the pyramid. In fact, they were enthusiastic enough to request additional time "sleeping-in". It behooves one to pay attention to a consensus statement of that type. Admittedly, there may be other reasons for the common group experience, but there is reason to continue to research the pyramid for no other reason than the strength of their common experience.

PYRAMID POWER II - CONCLUSIONS

V. CONCLUSIONS

As mentioned in the abstract, the purpose of this project was to demonstrate whether there was, in fact, <u>an effect</u> on a system due to placing that <u>system within a</u> pyramid - shaped structure for a length of time. It has indeed been shown that certain systems are definitely influenced by the pyramid structure:

> (1) The growth rate of black - eyed peas and lima beans is increased at a probability level of chance results of 0.06 and 0.28 respectively.
>
> (2) Raw hamburger meat is relatively preserved at a probability level of chance results of 0.001.
>
> (3) The growth rate of bacteria is decreased -- throat bacteria probability of chance results at 0.05 and yeast bacteria growth probability of chance results of approximately 0.10.
>
> (4) Sleeping within a pyramid structure tends to create an energized state upon waking.

Other less definitive results show that:

> (1) The weight of crystals

formed as a supersaturated solution cools is increased at a probability of chance results level of between 0.10 and 0.25.

(2) Dead organic matter is relatively preserved.

(3) The death or decay rate of a live plant isolated from water, light, and nutrient inputs is decreased.

APPENDIX

PYRAMID POWER II – DATA PROCESSING

DATA PROCESSING PLAN FOR PYRAMID DATA

This memorandum will define the capture of data from pyramid experiments on 80 column IBM cards and the subsequent processing of that data.

1. CARD FORMATS

A basic card format will be employed for all pyramid experiments. An illustration of this format is shown in TABLE A1. Provision is made in the format for a 20 column alphanumerical descriptor

PYRAMID POWER II - DATA PROCESSING

of the experiment in question. Provision has also been made for a two column numerical experiment identification code, a one column treatment identifier, a two column block identifier and a two column replicate identifier. The block and replicate identifiers should be in ascending order, starting with 01 and without missing numbers. Leading zeros should be filled in on the experiment, block and replicate codes. The remaining columns of the IBM card are allocated to various characteristics that are descriptive of the treatment-block-replicate.

In the case of the sleep experiment, the person will correspond to the block and

PYRAMID POWER II - DATA PROCESSING

there will be two treatments:

- Sleep within the pyramid (code 1)

- Sleep outside the pyramid (code 2)

For each person - treatment, there will be three replicates (generally) and five characteristics (parameters) recorded both before and after meditation for a total of eight parameters.

PYRAMID POWER II - DATA PROCESSING

2. DATA CAPTURE

In coding the data, each line of the coding form will be associated with the data from one block-treatment-replicate. The data for each experiment should be kept together as a unit on the coding forms.

3. DATA ANALYSIS

A separate analysis of variance will be performed for each characteristic in each experiment. This analysis will have the sources of variation listed.

PYRAMID POWER II - DATA PROCESSING

Analysis of Variance

Source	Degrees of Freedom
Blocks	$b - 1$
Treatments	$t - 1$
B × T	$(b - 1)(t - 1)$
Pooled error	$\sum_i \sum_j (r_{ij} - 1)$

PYRAMID POWER II - DATA PROCESSING

Where r_{ij} is the number of replicates in the j - th block of the i - th treatment. For some experiments, all r_{ij} will be 1 and hence the source of variation will not exist. The Datatext program employs the method of unweighted means (See Cochran and Cox).

In interpreting the analysis of variance, the treatments will be considered as fixed and the blocks will be considered as random. A consequence of these assumptions is that the appropriate mean square for testing for the existence of treatment effects is the B x T interaction mean square.

PYRAMID POWER II - DATA PROCESSING

TABLE A1

PYRAMID RESEARCH EXPERIMENT (SPRING 1975)
Card Formats for Experiments 01 through 07

Card Columns	Code	Field Description
1 - 20	A	Name of the Experiment
21 - 23		Blank
24 - 25	N	Experiment Number
26	N	Treatment
		Code — Interpretation
		1 — Pyramid
		2 — Cube
		3 — Uncovered
27 - 28	N	Block
29 - 30		Replicate
31 - 35		Characteristic*

*For all experiments with a single characteristic, the data is left justified. All decimals are punched.

PYRAMID POWER II - DATA PROCESSING

TABLE A2
PYRAMID RESEARCH EXPERIMENT (SPRING 1975)
Card Formats for Experiment 08

Card Columns	Code	Field Description
1 - 20	A	Name of Experiment
21 - 23		Blank
24 - 25	N	Experiment Number
26	N	Treatment

            Code   Interpretation
            1   Pyramid
            2   Cube
            3   Uncovered

Card Columns	Code	Field Description
27 - 28		Replicate
29 - 30	N	Block
		Characteristic*
		Before Sleep Data
31 - 35		Pulse
36 - 40		Temperature
41 - 45		GSR
		Blood Pressure
46 - 50		Systolic
51 - 55		Diastolic
		After Sleep Data
56 - 60		Pulse
61 - 65		Temperature
66 - 70		GSR
		Blood Pressure
71 - 75		Systolic
76 - 80		Diastolic

*The data in these fields is right justified and decimals are always punched.

PYRAMID POWER II - DATA PROCESSING

APPENDIX II

EXPERIMENTAL DATA
FROM THE PYRAMID EXPERIMENTS

PYRAMID POWER II - DATA PROCESSING

Experiment 01
Length of Sprouts in Inches of Black-eyed Peas

	Pyramid (1)	Treatment Cube (2)	Uncovered* (3)
Block 01	5.6	1.1	
	3.3	1.6	
	1.1	1.5	
	2.7	1.0	
	1.9	1.0	
	1.7	1.7	
	2.2	0.9	
	1.9	1.5	
	1.0	1.2	
	2.3	2.0	
	2.0	2.4	
	0.3	1.5	
	2.1	1.3	
	1.4		
	1.5		
Block 02	1.4	1.3	
	1.3	1.3	
	1.1	1.2	
	0.9	0.8	
	1.0	0.5	
	0.8	1.2	
	1.5	1.6	
	1.3	1.3	
	1.5	1.2	
	1.3	1.0	
	1.1	1.6	
	1.0	1.1	
	1.5	2.0	
	--	1.0	

* There was no uncovered treatment in this experiment

PYRAMID POWER II - DATA PROCESSING

Experiment 01
Continued

	Pyramid (1)	Treatment Cube (2)	Uncovered* (3)
Block 03	0.51	0.39	
	1.85	0.20	
	0.43	1.06	
	0.71	0.63	
	0.98	0.47	
	0.94	---	
	0.52	---	
Block 04	3.15	1.38	
	2.17	0.20	
	0.39	1.57	
	1.77	1.18	
	1.57	0.79	
	2.17	0.20	
	1.77	0.39	
	2.56	0.79	
	2.56	1.18	
	1.38	---	
	2.36	---	
	1.57	---	

*There was no uncovered treatment in this experiment.

PYRAMID POWER II - DATA PROCESSING

Experiment 01
Continued

	Pyramid (1)	Treatment Cube (2)	Uncovered* (3)
Block 05	2.8	2.2	
	2.5	2.0	
	2.9	1.1	
	2.4	2.0	
	2.0	2.2	
	2.6	1.4	
	2.3	2.2	
	2.4	1.9	
	1.7	2.1	
	1.7	2.0	
	2.0	1.1	
	2.2	2.0	
	3.6	1.7	
	1.7	1.4	
	1.7	1.5	
	1.4	2.5	
Block 06	2.2	2.5	
	2.1	3.1	
	2.6	2.8	
	2.8	2.5	
	2.0	2.6	
	2.9	2.1	
	2.7	3.1	
	3.2	1.7	
	2.4	1.8	
	2.0	2.1	
	3.3	2.0	
	2.2	1.8	
	2.4	2.6	
	---	1.7	
	---	2.6	
	---	2.6	
	---	2.0	
	---	1.9	

PYRAMID POWER II - DATA PROCESSING

Experiment 02
Length of Sprouts in Inches of Lima Beans

	Treatment		
	Pyramid	Cube	Uncovered*
	(1)	(2)	(3)
Block 01	1.26	0.39	
	1.46	0.20	
	1.34	1.06	
	0.51	0.63	
	1.85	0.47	
	0.43	----	
	0.71	----	
	0.98	----	
	0.94	----	
	0.51	----	
	0.83	----	
	1.61	----	
Block 02	3.93	1.18	
	0.39	0.79	
	2.36	0.79	
	0.39	1.18	
	2.36	0.79	
	2.76	0.60	
	3.15	0.39	
	2.36	1.18	
	1.57	0.79	
	0.79	0.39	
	4.33	----	
	2.36	----	

* There was no uncovered treatment in this experiment

PYRAMID POWER II - DATA PROCESSING

Experiment 03
Freshness of Meat (Hamburger) on a 5-point Scale

	Treatment		
	Pyramid	Cube	Uncovered*
	(1)	(2)	(3)
Block 01	2	5	5
	2	4	5
	4	5	5
	3	5	4
	3	5	5
	3	5	5
	3	5	5
Block 02	2	5	4
	2	5	5
	2	4	4
	2	4	4
	3	4	5
	3	5	5
	4	4	5
Block 03	1	4	5
	1	5	4
	1	5	4
	1	4	4
	1	5	4
	2	5	5
	1	5	5

PYRAMID POWER II - DATA PROCESSING

Experiment 03
Continued

	Pyramid (1)	Treatment Cube (2)	Uncovered* (3)
Block 04	2	5	4
	2	5	5
	1	5	5
	2	–	4
	2	–	4
	3	–	5
	2	–	5
Block 05	2	5	5
	3	5	5
	3	5	5
Block 06	2	4	4
	2	3	5
	2	4	5
Block 07	2	4	5
	2	5	4
	3	5	4
Block 08	2	5	4
	3	5	5
	3	4	5

PYRAMID POWER II - DATA PROCESSING

Experiment 04

Normalized Area of Throat Bacteria

	Treatment		
	Pyramid	Cube	Uncovered*
	(1)	(2)	(3)
Block 01	243	471	531
Block 02	133	343	351
Block 03	042	122	058
Block 04	106	137	099
Block 05	099	157	263

PYRAMID POWER II - DATA PROCESSING

Experiment 05

<u>Sum of Root Lengths in Inches of Rootings</u>

	Treatment		
	Pyramid	Cube	Uncovered*
	(1)	(2)	(3)
Block 01	00.6	26.2	39.2
Block 02	00.6	40.6	26.4

PYRAMID POWER II - DATA PROCESSING

Experiment 06
Weight of Crystal Formed in Grams

	Pyramid (1)	Treatment Cube (2)	Uncovered* (3)
Block 01	10.1	11.7	12.3
	13.3	13.5	----
	13.8	----	----
Block 02	17.2	17.1	14.4
	18.4	15.7	----
	18.2	----	----
Block 03	30.5	30.0	27.8
	30.1	31.5	----
	29.8	----	----

PYRAMID POWER II - DATA PROCESSING

Experiment 07
Normalized Area of Yeast Bacteria with Respect to
Mean Area of Uncovered Treatment

	Pyramid (1)	Treatment Cube (2)	Uncovered* (3)
Block 01	0.58 0.50	1.19	0.54
Block 02	0.38 0.50	0.54	1.46
Block 03	0.50	0.56	1.18
Block 04	0.59	0.75	0.87
Block 05	0.46	0.96	0.84
Block 06	0.51	1.15	1.10
Treatment Means	0.50	0.86	1.00

PYRAMID POWER II - DATA PROCESSING

Experiment 08
Experimental Data for the Sleep Experiment

	Before Sleep					After Sleep				
				Blood Pressure					Blood Pressure	
Pulse (Cnt)	Temp F	GSR (Norm)	Systolic mm of Hg	Diastolic mm of Hg	Pulse (Cnt)	Temp F	GSR (Norm)	Systolic mm of Hg	Diastolic mm of Hg	

BLOCK 01

Treatment 01 (Sleeping Inside)

Pulse (Cnt)	Temp F	GSR (Norm)	Systolic	Diastolic	Pulse (Cnt)	Temp F	GSR (Norm)	Systolic	Diastolic
64	98.4	2.1	100	063	62	98.4	4.2	116	078
76	97.8	2.1	110	078	72	97.4	2.7	098	062
62	98.0	2.2	105	065	60	97.1	2.2	104	076
66	97.6	2.5	104	070	66	97.6	2.9	106	072

Treatment 02 (Sleeping Outside)

Pulse (Cnt)	Temp F	GSR (Norm)	Systolic	Diastolic	Pulse (Cnt)	Temp F	GSR (Norm)	Systolic	Diastolic
60	98.2	1.9	116	070	60	97.2	6.5	108	068
60	96.3	3.0	115	080	66	97.7	2.0	112	072

PYRAMID POWER II - DATA PROCESSING

Experiment 08
Continued

Before Sleep					After Sleep				
Pulse (Cnt)	Temp F	GSR (Norm)	Blood Pressure Systolic Diastolic mm of Hg		Pulse (Cnt)	Temp F	GSR (Norm)	Blood Pressure Systolic Diastolic mm of Hg	
BLOCK 02									
Treatment 01 (Sleeping Inside Pyramid)									
90	98.6	1.8	115	070	70	97.6	3.0	130	085
64	97.8	2.1	128	094	70	97.6	2.5	124	084
80	98.2	2.1	145	080	70	96.6	2.3	122	090
60	97.8	2.2	128	073	70	98.7	2.5	120	078
Treatment 02 (Sleeping Outside)									
68	98.0	1.9	135	070	50	97.7	6.5	112	072
64	98.1	2.1	118	079	62	97.2	2.2	108	065

PYRAMID POWER II - DATA PROCESSING

Experiment 08
Continued

	Before Sleep					After Sleep			
			Blood Pressure					Blood Pressure	
Pulse (Cnt)	Temp F	GSR (Norm)	Systolic mm of Hg	Diastolic mm of Hg	Pulse (Cnt)	Temp F	GSR (Norm)	Systolic mm of Hg	Diastolic mm of Hg
				BLOCK 03					
			Treatment 01 (Sleeping Inside)						
82	98.2	1.3	135	080	86	98.6	2.4	120	070
80	97.6	2.0	152	098	68	96.8	2.3	122	080
92	97.0	2.3	135	090	80	97.0	2.3	120	080
80	96.2	2.3	130	080	80	96.2	3.0	134	092
			Treatment 02 (Sleeping Outside)						
72	97.8	1.8	130	090	68	96.8	7.0	124	080
38	98.6	1.7	140	085	76	97.2	2.0	120	072

80

PYRAMID POWER II - DATA PROCESSING

Experiment 08
Continued

	Before Sleep				After Sleep			
			Blood Pressure				Blood Pressure	
Pulse (Cnt)	Temp F	GSR (Norm)	Systolic Diastolic mm of Hg		Pulse (Cnt)	Temp F	GSR (Norm)	Systolic Diastolic mm of Hg	

BLOCK 04

Treatment 01 (Sleeping Inside Pyramid)

84	98.7	2.0	150	080	80	97.4	1.8	135	080
90	98.6	1.9	158	092	70	98.0	2.2	148	098
96	98.6	1.7	160	090	66	98.0	2.1	143	098
72	98.2	1.9	128	076	72	96.4	2.2	128	088

Treatment 02 (Sleeping Outside)

| 86 | 98.9 | 1.8 | 160 | 100 | 60 | 96.3 | 2.3 | 142 | 084 |
| 78 | 97.6 | 1.9 | 135 | 100 | 72 | 97.3 | 1.9 | 140 | 100 |

PYRAMID POWER II - DATA PROCESSING

Experiment 08
Continued

	Before Sleep					After Sleep			
			Blood Pressure					Blood Pressure	
Pulse (Cnt)	Temp F	GSR (Norm)	Systolic mm of Hg	Diastolic mm of Hg	Pulse (Cnt)	Temp F	GSR (Norm)	Systolic mm of Hg	Diastolic mm of Hg

BLOCK 05
Treatment 01 (Sleeping Inside)

Pulse (Cnt)	Temp F	GSR (Norm)	Systolic	Diastolic	Pulse (Cnt)	Temp F	GSR (Norm)	Systolic	Diastolic
88	99.8	2.1	135	070	88	97.0	3.0	118	086
80	98.6	1.9	134	080	72	98.6	1.9	128	074
98	99.2	1.6	120	070	88	98.0	2.1	118	082
86	98.5	1.9	098	062	80	97.2	1.9	122	080

Treatment 02 (Sleeping Outside)

Pulse (Cnt)	Temp F	GSR (Norm)	Systolic	Diastolic	Pulse (Cnt)	Temp F	GSR (Norm)	Systolic	Diastolic
94	99.3	1.4	118	080	66	98.1	2.1	120	085
68	98.6	1.2	130	060	72	98.5	1.4	110	070

PYRAMID POWER II - DATA PROCESSING

Experiment 08
Continued

	Before Sleep					After Sleep			
Pulse (Cnt)	Temp F	GSR (Norm)	Blood Pressure Systolic mm of Hg	Blood Pressure Diastolic mm of Hg	Pulse (Cnt)	Temp F	GSR (Norm)	Blood Pressure Systolic mm of Hg	Blood Pressure Diastolic mm of Hg
BLOCK 06									
Treatment 01 (Sleeping Inside Pyramid)									
90	98.2	2.2	115	075	90	97.7	2.0	116	078
90	98.2	1.9	102	068	66	98.0	2.2	100	072
84	98.5	1.8	120	080	78	98.2	2.2	105	070
62	98.0	2.0	123	085	80	97.6	4.0	118	084
Treatment 02 (Sleeping Outside)									
72	98.7	2.0	108	068	64	97.7	2.2	090	058
80	97.6	4.3	115	068	54	97.7	2.2	118	078

PYRAMID POWER II - STATISTICAL HANDLING

<u>EXAMPLES OF</u>

<u>STATISTICAL CALCULATIONS</u>

<u>EXPERIMENTS 04 THROUGH 06</u>

PYRAMID POWER II - STATISTICAL HANDLING

EXPERIMENT 04
Normalized Area of Throat Bacteria

Block	Pyramid (1)	Cube (2)	Uncovered (3)	Totals	Means
01	243	471	531	1245	415.0
02	133	343	351	827	275.7
03	042	122	058	222	74.0
04	106	137	099	342	114.0
05	099	157	263	519	173.0
Totals	623	1220	1302	3155	
Means	124.6	244.0	260.4	210.33	

Correction Factors = $(3155)^2/15$ = 663601

Block Uncor ss = $\dfrac{(1245)^2+\ldots(519)^2}{3}$

Treat Uncor ss = $\dfrac{(623)^2+\ldots(1302)^2}{5}$ = 714347

Total Uncor ss = $(243)^2+\ldots(263)^2$ = 984827

ANALYSIS OF VARIANCE

Source	Degrees of Freedom	Sum of Squares	Mean Square	F
Blocks	4	226253	56563	10.23
Treatments	2	50746	25373	4.59*
Error	8	44227	5528	
Total	14	321226		

* The hypothesis that the three treatments have equal effectiveness is rejected at the 5% significance level.

PYRAMID POWER II

EXPERIMENT 05
Sum of Root Lengths in Inches of Rootings

Block	Pyramid (1)	Cube (2)	Uncovered (3)	Totals
01	0.6	26.2	39.2	66.0
02	0.6	40.6	26.4	67.6
Totals	1.2	66.8	65.6	133.6
Means	0.6	33.4	32.8	66.8

Correction factor = $(133.6)^2/4$ = 2974.83

Corrected Block ss = $\dfrac{(66.0 - 67.2)^2}{6}$

Treat Uncor ss = $\dfrac{(1.2)^2 + \ldots (32.8)^2}{2}$

Total Uncor ss = $(0.6)^2 + \ldots (26.4)^2$ = 4569.12

Cube vs Uncovered = $\dfrac{(33.4 - 32.8)^2}{4}$

ANALYSIS OF VARIANCE

	Degrees of Freedom	Sum of Squares	Mean Square	F
Blocks	1	0.43	0.43	
Treatments	2	1408.69	704.35	7.61
Pyramid vs others	1	0.09		15.21
Cube vs uncovered	1	1408.60	1408.60	
Error	2	185.17	92.59	
TOTAL	5	1594.29		

The hypothesis that the pyramid has the same effect as the other two treatments is rejected at the 10% significance level.

EXPERIMENT 06
Weight of Crystals Formed in Grams
Basic Data

	Pyramid (1)	Cube (2)	Uncovered (3)
Block			
01	10.1	11.7	12.3
	13.3	13.5	
	13.8		
02	17.2	17.1	14.4
	18.4	15.7	
	18.2		
03	30.5	30.0	27.8
	30.1	31.5	
	29.8		

Table of Totals

Block	Treatment		
	(1)	(2)	(3)
01	37.2	25.2	12.3
02	53.8	32.8	14.4
03	90.4	61.5	27.8

Table of Means

Block	Treatment			Totals
	(1)	(2)	(3)	
01	12.4	12.6	12.3	37.3
02	17.9	16.4	14.4	48.7
03	30.1	30.8	27.8	88.7
Totals	60.4	59.8	54.5	174.7
Means	20.1	19.9	18.2	19.4

PYRAMID POWER II

EXPERIMENT 06
continued

Correction Factor = $\dfrac{(174.7)^2}{9}$ = 3391.12

Block Uncor ss = = 3398.15

Total Uncor ss = = 3888.03

ANALYSIS OF VARIANCE

Source	Degrees of Freedom	Sum of Squares	Mean Square	F	Probability
Blocks	2	485.77	242.9		
Treatments	2	7.03	3.5	3.2	
Error	4	4.11	1.1		0.10 < P < 0.25
Total	8	496.91			

PYRAMID POWER II - MEDITATION

EXPERIMENT 13: MEDITATION

The purpose of the Meditation Experiment was to determine whether the process of meditation is affected by performing the meditation within a reflective pyramidal structure as compared to meditation outside of that structure. The working definition of meditation used for the purpose of this experiment was: experiencing fifteen minutes of relaxation in a quiet dark room while listening to a tape recording of a metronome sound pattern. The sound pattern consisted of a superposition of two vibratory patterns, (1) a constant frequency clicking at 10 cycles per second and (2) a decreasing frequency clicking initially at 72 cycles per minute and decreasing linearly to 45 cycles per minute over fifteen minutes.

The measurements consisted of four physiological variables measured before and after each meditation period and subjective observations as to the depth of meditation and any unusual meditative experiences. The physiological measurements taken were: (1) systolic and diastolic blood pressure (mm of Hg); (2) Galvanic Skin Resistance (GSR), a nondimensional number characterizing the null balance of the GSR resistance bridge circuit, (3) Kirlian Electro-photography of the fingertips of each hand; and (4) the dominant brain wave frequency (cycles per second) and it amplitude in microvolts.

PYRAMID POWER II - MEDITATION

The pyramid treatment structure was a square-based pyramid with a six foot base and a height of 3.81 feet. The faces of the pyramid were constructed of a mylar sheet with aluminum foil bonded on the inner surface; this material has a high percentage of reflectivity for electromagnetic energies from the Ultra-Violet wavelengths through the far Infrared spectrum. The human body emits coherent energy in the ranges of 350 nanometers through 10 micrometers (editor's note: see Electromagnetic Bio-Information by F.A. Popp). Hence energy given off from the body will be reflected by the surface. The structural configuration was designed to reflect this energy to a focal point approximately 1.3 feet above the base and directly under the apex of the pyramid.

There were thirteen subjects, each of whom meditated six times during the course of the five-week experiment --- three times inside the pyramid structure and three times outside the structure. The order of meditation with respect to inside or outside was randomly assigned for each subject.

The results of the experiment were evaluated statistically using analysis of variance. Differences between 'Before' and 'After' meditation variable values were calculated and these differences were compared for pyramid structure treatment meditation sessions vs. non-pyramid structure treatment sessions.

PYRAMID POWER II - MEDITATION

The following table shows the Summary of Results for the Meditation Experiment. Analysis of the mean values of the Before Meditation data for the Pyramid Treatment and the Non-Pyramid Treatment shows that initially the variable values were essentially equal. However, the amplitude of the dominant frequency for the pyramid treatment before meditation is 10% higher than that for the non-pyramid and this fact is inexplicable. There is, in fact, a difference in the After-Meditation Data which indicates that the pyramid treatment <u>does</u> <u>affect</u> at least one of the measured physiological variables.

Analysis of the difference between the After Meditation and the Before Meditation variable shows that (1) the pyramid treatment reduces the <u>systolic blood pressure</u> by 8.6 mm of Hg as compared to the non-pyramid treatment difference of 3.5 mm of Hg at a probability level of 0.027. (2) The <u>diastolic blood pressure</u> decreases by essentially the same amount for each treatment. (3) Meditation, as expected, increases the <u>GSR</u> <u>measurement</u> in both treatments indicating a reduction of tension. The pyramid treatment increased the GSR by an amount slightly greater than the non-pyramid treatment; however, the difference is not significant statistically (0.388). (5) The <u>amplitude of the dominant frequency</u> decreases during meditation and again there is no statistical confirmation of a real difference between treatments.

PYRAMID POWER II - MEDITATION

The amplitude during pyramid treatment decreases more than the non-pyramid treatment but to a probability level of 0.285.

There is some indication that meditation within a pyramid tent has an effect physiologically when compared to meditation outside of a pyramid tent. The results of this preliminary experiment show that a greater systolic blood pressure drop is experienced, a slightly greater decrease in tension (as measured by an increase in the GSR), and a different alteration in dominant brain-wave frequency and amplitude is effected by meditation in a pyramid tent. Nevertheless, it should be noted that the results are not statistically significant at the 5% probability level.

PYRAMID MEDITATION AND SLEEP TENT

Two happy subjects peer out from the interior of the aluminized pyramid tent used in the meditation and sleep experiments.

MEDICAL ELECTRONICS EQUIPMENT

This photograph illustrates a portion of the medical electronics equipment used to monitor physiological parameters during the pyramid meditation and sleep experiments.

PYRAMID POWER II – MEDITATION

TABLE A3

Experiment 13

SUMMARY OF RESULTS FOR MEDITATION EXPERIMENT

(Number of Blocks = 13)

Variable	Computer Variable	Treatment Means Inside Tent	Outside Tent	Probability*
After Minus Before				
Systolic (mm of Hg)	DIFF (1)	-8.58	-3.54	0.027
Diastolic (mm of Hg)	DIFF (2)	-1.44	-1.03	>0.500
GSR (normalized no.)	DIFF (3)	+0.45	-0.02	0.289
Dominant Brain Wave:				
Freq. (Hz per sec.)	DIFF (4)	+0.89	+0.56	0.383
Amp. (microvolts)	DIFF (5)	-8.72	-4.62	0.285
Ratio of After to Before				
Systolic (mm of Hg)	RAT (1)	0.929	0.97	0.035
Diastolic (mm of Hg)	RAT (2)	0.985	0.989	>0.500
GSR (normalized no.)	RAT (3)	1.165	1.058	0.277
Dominant Brain Wave:				
Freq. (Hz per sec)	RAT (4)	1.09	1.06	0.355
Amp. (microvolts)	RAT (5)	0.848	0.94	0.320

(continued)

* Probability of observing by chance alone

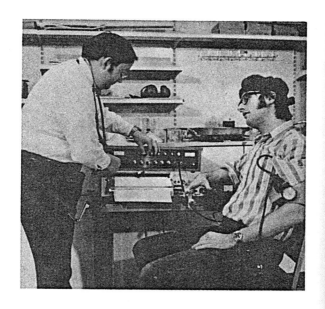

GSR AND BLOOD PRESSURE TEST

The subject in the above photograph is receiving a GSR (Galvanic Skin Resistance) and Blood Pressure test. The subject has just meditated in the aluminized pyramid tent structure.

TABLE A3 continued

Experiment 13

SUMMARY OF RESULTS FOR MEDITATION EXPERIMENT

(Number of Blocks = 13)

Variable	Computer Variable	Treatment Means		Outside Tent Probability*
		Inside Tent	Outside Tent	
Before Meditation Data				
Systolic (mm of Hg)	CHAR 1 (1)	116.23	111.82	0.043
Diastolic (mm of Hg)	CHAR 1 (2)	73.62	73.33	>0.500
GSR (normalized no.)	CHAR 1 (3)	3.71	3.43	0.384
Dominant Brain Wave:				
Freq. (Hz per sec.)	CHAR 1 (4)	11.87	11.67	>0.500
Amp. (microvolts)	CHAR 1 (5)	45.95	40.97	0.092
After Meditation Data				
Systolic (mm of Hg)	CHAR 2 (1)	107.65	108.28	>0.500
Diastolic (mm of Hg)	CHAR 2 (2)	72.18	72.31	>0.500
GSR (normalized no.)	CHAR 2 (3)	4.15	3.41	0.084
Dominant Brain Wave:				
Freq. (Hz per sec)	CHAR 2 (4)	12.76	12.23	0.256
Amp. (microvolts)	CHAR 2 (5)	37.23	36.36	>0.500

* Probability of observing by chance alone

PYRAMID POWER II - MEDITATION

SUBJECTIVE OBSERVATIONS

Two subjective experiences are worth noting. In the first instance, a female volunteer who was aware of her healing energies (a build-up and discharge of energy from herself to another person during a particular state of consciousness) meditated in the pyramid structure. During the first thirty seconds of meditation, she noticed a definite build-up of her healing energies. This build-up continued for the duration of the fifteen minute session and caused her considerable mental anguish as there was no apparent outlet for the energies. At the end of the fifteen minutes, she was visibly very upset from this unexpected build-up of energies and the lack of an outlet channel to discharge it. This experience is similar to many others reported in the literature regarding a build-up and enhancement of psychic energy due to meditation in a pyramid structure in contrast to meditation outside of such a structure.

The second instance was experienced by the experimenter, Lee Bristol. He was meditating outside of the pyramid structure while a female volunteer was meditating inside the pyramid. At the end of the metronome tape recording, which signified the end of the meditation session, he continued meditating as many vivid visual images were passing through his consciousness. The last image held was

PYRAMID POWER II - MEDITATION

that of the female volunteer, who had been meditating in the pyramid, coming into Bristol's 'field of vision' and waving her hands in front of his face to attract his attention. This particular type of movement was known to be characteristic of the girl volunteer and, as he recognized it, he stopped his meditation. Upon asking the volunteer what she had just been thinking, she replied: "I was thinking, Lee, the session is over, let's stop it." It appears that the thought form of the volunteer manifested itself in Bristol's consciousness, not only carrying information to be communicated but also carrying information as to the identity of the sender as conveyed by characteristic mannerisms seen in Bristol's visualization. It should be noted that the female volunteer was still in the pyramid when the thought transmission occurred.

While no attempt is made to attribute this telepathic communication to the pyramid structure, nevertheless, a similar spontaneous experience was observed at another time under similar conditions with a different volunteer. Again a message to stop the meditation was communicated in a form which also included mannerisms characteristic of that particular sender. The sender was asked her thoughts at the time of the communication and she replied: "I was thinking, Lee, it is time to stop this meditation."

NOTES FROM KAREL DRBAL
THE PYRAMID RAZOR BLADE REGENERATOR

PYRAMID POWER II

LETTERS FROM KAREL DRBAL

Karel Drbal is the Czechoslavakian radio engineer who obtained a patent on a pyramid model razor blade 'sharpener' or regenerator. He spent over 10 years arguing his case on the patent before the examiner approved it for release. During this period he had to prove that the pyramid did indeed have a measurable effect before the patent would be approved for issue. Mr. Drbal has carried on an active correspondence with Dr Carl Schleicher of Mankind Research for a number of years. It is with his kind permission that we duplicate a number of his letters. We are duplicating the language of his letters as they were written, without correction of English grammer in order to fully communicate his ideas on this subject area.

PYRAMID POWER II

KAREL DRBAL, ENGINEER

Translation of the Main Passages of the Czech Patent Specification No. 91304

Filed: November 4, 1949

Issued: August 15, 1959

Ten years later – that means the examination of the patent application lasted ten years.

A PROCEDURE FOR MAINTAINING THE SHARPNESS OF RAZOR BLADES AND RAZORS

This invention concerns a procedure for maintaining the sharpness of razor blades and straight razors without using any artificial source of energy. In this procedure of acting on the sharpness of razor blades and straight razors we do not, according to this invention, use any mechanical, thermic, or chemical sharpening devices or electric devices from an artificial electric source.

Yet it is customary for the sharpening of blunt razor blades to make use of mechanical grindstone devices, where the blunted edge is sharpened by mechanical action of the grindstone, which results in the loss of blade material by the sharpening procedure. Furthermore, it is well known that in order to ameliorate the sharpness of razor blades and straight razors by using an artificial magnetic

PYRAMID POWER II

field, the blade is placed so, that principally the edge lies in the magnetic force line direction.

According to the invention, the razor blade is placed in the direction of the Earth's magnetic field under a hollow pyramid, made of dielectric material, e.g., cardboard, cardboard impregnated with paraffin, or any plastic. The pyramid has in its base side a hole for introducing the blade. This hole can have a square form, round, or oval shape.

As the best shape for the pyramid a quadrangle type is recommended with a square base, and it is stated, that advantageously the baseline dimension is found by multiplying the height of the pyramid with the number of Ludolff. So e.g., if the height of the pyramid is 10 cm, the baseline of the quadrangle is 15.7 cm.

The razor blade or straight razor is placed on a support also made of dielectric material -- the same as used for the pyramid or other, e.g., cork, wood, stone-material, paper, paper impregnated with paraffin and the like -- and the height of this support is placed at the 1/5 - 1/3 of the pyramid height. This support must be made so that the edges of the blade are free in the air, and do not lie or touch the support material.

In order to ameliorate the action

PYRAMID POWER II

of the pyramid on the blade, it is recommended to place the blade with its longitudinal axis in the direction of the Earth's magnetic field, North-South. Such position is ameliorating the function of this device, but is not indispensable for using the pyramid according to this invention. The walls of the pyramid can be advantageously placed in the four compass-directions: North, South, East, and West.

It is recommended to leave a new blade one to two weeks in the pyramid before the beginning of the shaving and replace it there immediately after shaving. It is not admissible to place in the pyramid-device an **old blunted blade.** By using this device according to the specifications, it was possible to make 1,778 shaves with only 16 'Dukat-Zlato' blades, which means an average of 111 shaves with the same blade. The minimum in this experimental observation was 50, and the maximum was 200 shaves with the same blade.

(This is the text of pages 1 and 2, up to line 51 of the patent description.)

Very important is the last paragraph of the patent description, lines 61 - 67:

> This invention was especially tested for a specific pyramid-shape device, but is

PYRAMID POWER II

NOT RESTRICTED ONLY TO THIS ONE SPECIFIC FORM, which means that it can also be valuable for <u>other</u> geometrical shapes of dielectric material, used in the manner described in this invention with the following definition:

In the space, enclosed by this form, will start an automatic regenerative process concerning the fine structure of the razor blade edge and produced only by the specific cavity according to the invention. (That means, that the excitation of this cavity is only produced by the surrounding cosmic and terrestrial fields.) This process acting on the razor-blade edge by producing a diminution of the number of inner disturbances/calculated dislocations, provoked by the shaving process, in the lattice-bands of the micro-crystalline structure of the edge shape, that have as a result a <u>regeneration</u> of the edge material and its inner fine, crystalline structure; regeneration that causes a RENOVATION of the mechanical and physical properties of the razor-blade edge, removing the

PYRAMID POWER II

fatigue of material issued from the shaving action (and only all this, if the crystal lattice disturbances are of an elastic type and not of a definitive type, e.g., mechanical destruction of the edge).

REMARKS

To the text of this important last paragraph of the patent specification description, I have added some explanations for further understanding.

This razor-blade pyramid is today not produced or marketed in Czechoslavakia because the blade manufacturer with foreign license has no interest, to see someone use his 'Astra' blade 100 or 200 times, instead of the guaranteed 8 to 12 times.

There is no direct statement that the pyramid-device is a sharpener because it is only a regenerating device without any mechanical action. Only with micro-wave resonant action on the microstructure of the blade edge crystals, if the deformation of the lattice is not of a definitive type and if by inner displacement in the crystal lattice levels a regeneration by very weak micro-wave resonance in a pyramid resonant cavity, is this action possible. This is the true 'mystery' of this little pyramid. The first condition before all others, is the use of FIRST CLASS QUALITY steel edge blades. For bad, old fashioned or deteriorated blades, this patented device naturally can not give any results.

During the ten years of examining the patent, the owner had elaborated for the Patent Office a working theory of his device as a micro-wave horn or micro-wave dielectric antenna fed through the dielectric walls with cosmic energy and acting by resonance on the blade edge. The mean source is supposedly the Sun with its very broad micro-wave spectrum and it was calculated, that the energy of this little dielectric antenna can be sufficient, because 1 - 1.5 eV (electron volts) is only needed for acting on the micro-crystalline grid of the blade edge.

The best recommendation during this examination was the fact, that the Chief Examiner, Engineer Vrecion, an excellent metallurgist, asked me to make for him a razor-blade pyramid (8 cm high, baseline 12.5 cm) which has functioned very well all this time. At first, it was for him as he told me, the greatest surprise of his life. He was later himself the best defender, when other examiners would not give me this patent and made a great number of objections, which since then I have explained. This is why the patent examination lasted approximately 10 years, much longer than customary.

In the text of the patent-specification is a very good account

PYRAMID POWER II

of the whole story of this unusual patent
and its functioning.

            Karel Drbal
         February 25, 1974

PYRAMID POWER II

KAREL DRBAL'S IN DEPTH DESCRIPTION OF THE PYRAMID RAZOR BLADE REGENERATOR

The little Cheops-Pyramid cardboard model with its strange abilities is not a 'magical' but a VHF (Very High Frequency) or UHF (Ultra High Frequency) device, e.g., multiplying by 10 or more times the possibility of using one razor blade. This is a short story about this strange invention, known in many countries in Europe and also Overseas.

I speak about patent number 91304, more or less indicating that the cavity of a little cardboard model of the Great Pyramid of Cheops can 'work' on a steel edge!

At the first sight of the patent specification there is something very unusual in the examination time of this patent: filed in 1949, patented only in 1959! The examining commission took just 10 years! - something very astonishing, because the normal time of examination to patent is between 1 to 3 years. In the course of this 10 years I was forded to search new and others scientific arguments and explanations, to demonstrate how this device without any evident source of energy, being of extreme simplicity, can 'work' on the steel-edge of a razor blade,

PYRAMID POWER II

fatigued by shaving.

When I demanded this patent, it was more of a joke for me and my friends, also radio-engineers, who encouraged me to do it and to see, how the Patent Office would react on a 'Pharoah's Shaving Device'. I must hereby emphasize, that a great number of these friends and of course myself have obtained more than 100 shaves with only one razor blade, using the 'pyramid regenerator', persuaded, that this strange device really works.

But this was the one side of the story, and the other side was to persuade the patent examiners: (1) that it works, and (2) how it works, which was just the greatest difficulty.

This 'how it works' forced me for 10 years to study all the possibilities of micro-wave, cosmic, and telluric relations between the resonant cavity of a Cheops-pyramid model, constructed of dielectric material (cardboard or other), and the 'working' of the crystalline structure of the edge of the razor blade, to study also, how the very low Earth magnetic field is engaged in this action, because one of the conditions of the patent is that the edge of the blade must lie in a longitudinal axis in the direction of the horizontal component of the Earth's magnetic field.

It was for me of greatest

importance, that I was employed as a radio engineer in a great Radio Institute of Research, where I could easily study all the necessary technical literature of the entire world, and so step by step, in the 10 years' fighting with the patent examiners, I was able to realize a theory (or hypothesis) about the energizing of the resonant cavity of the little pyramid model by cosmic microwaves (principally from the Sun) with the help of the concentrating Earth's magnetic field, - about the technical possibility of such energy feeding of the pyramid, and I was in this way able to persuade the examiners, that indeed the Pharoah Khufu (Kheops or Cheops) has nothing in common with the razor blade edges, and that the whole thing is not nonsense.

It was during the 10 years of patent examination that I constructed a little pyramid model in cardboard (8 cm high, baseline 12.5 cm, type Cheops) for the Chief Examiner, an excellent metallurgical specialist. The pyramid worked to his satisfaction during this 10 years and obliged him to defend my invention in the examining commission, gives him the possibility of proof by his own experience that it is not a mystification, but only a strange fact. Without the help of this honest examiner I am sure, that the strange patent No. 91304 would not exist.

The patent is conceived for the

'Cheops type', where the baseline can be easily calculated by multiplying the height of the pyramid by pi/2 that is 1.57079, what is exactly specified in the patent description, without limiting the invention only on this specific form, because I have found by an extremely great number of different experiences, that also other pyramidal shapes (types) are able to 'work' on the razor blade edge in the same manner as the Cheops type. I have specified this possibility in the patent description. The text, concerning this fact, is inserted in the last paragraph of the patent description, which also indicates why (with regard to my hypothesis) the pyramid-model cavity works or is supposed to work on the fine crystalline structure of the edge.

A necessary supposition is that the blade steel is of the very first quality, so that the deformation of the microstructure blade-edge, affected by the multiple shavings, doesn't be of a definitive (plastic) character, but of an elastic character.

The pyramid, type Cheops, or other forms, or any appropriate resonator for the same use, has only to produce an acceleration in the removing of the elastic deformation to the original (or nearly original) state of the edge, acceleration, which instead of normally 15 -30 days (without a regenerating device) is improved in only 24 hours! This is the real secret of the pyramid resonant cavity action on

PYRAMID POWER II

the edge of razor blades or straight razors.

But there is one more interesting effect, discovered by the scientist Professor Carl Benedicks from Stockholm (see: Metallkundliche Berichte, Verlag-Technik, Berlin 1952, Tome II. "Aenderung der Festigkeit von Mettallen und Nichmetallen durch eine benetzende Flussigkeit") -the so called Flussigkeitseffekt, the 'liquid strain effect' which produces in steel material a non-corrosive but steel hardness reducing action; e.g., water acting on steel can reduce its firmness by as much as 22%! This fact is very inhibitive to the micro-edge cavities, from where it is practically impossible to drive out the harmful water dipole molecules.

The pyramid (or other appropriate resonant cavity) is the only device that will do (if fed by micro-waves) this 'helpful job' for the razor blade-edge crystal gaps, to drive out the water molecule dipoles by resonant action on that dipole. We can symbolically say that the edge of the razor blade is being dehydrated by this action.

That such an action on the dipole molecules of water is possible in a resonant cavity, fed with appropriate micro-wave energy, was provided by scientists Born and Lertes (see Archiv der Electrischen Uebertragung, 1950, Heft 1,

PYRAMID POWER II

S.33 - 35, "Der Born-Lertessche Drehfeldeffekt in Dipolflussigkeiten im Gebiet der Zentimeterwellen"). It was found, that the micro waves of centimeter-wavelength and their harmonics can produce an accelerated rotation of the water dipole molecules, and this effect can have as a result the "driving out" of water dipole molecules from the smallest cavities, projecting it into the open air. This is exactly the process known as electromagnetic dehydration.

There is another question: why must the pyramid models be made from dielectric material? The answer is simple: because the micro-waves can penetrate through this material and energize (feed) the resonant cavity. This is a very old finding (see Journal of Applied Physics, Vol. 10, June 1959, pg. 391 - 398: Richtmyer, R.D., Stanford University, "Dielectric Resonators").

Still one more question: In microwave technology, the microwave resonator should be fed by some little antenna or by a coupling hole. Our pyramid can be without any hole, without affecting its function. We have explained that microwaves can penetrate through dielectric material (if microwaves are really in action here). This was confirmed experimentally by microwave technicians, as e.g.: Electronique, Revue Technique Electronique No. 118, September 1956, pg 10 - 13, Henry Copin, Inginieur au Service des

PYRAMID POWER II

Transmissions Militaires, "De l'existence Possible D'ondes Stationaires dans les Cellules Vivantes" (Possibility of Standing Wave Excitation in the Living Cells). This author postulates each living cell to be a microwave resonator and as a radio engineer explains the mechanism of cavity excitation with its surrounding walls consisting of dielectric or semi-conductor material.

The objection of the examiners, that the pyramid shape is normally not found in microwave devices was easy to reject with regard to some literature I presented. For example: Zeitschrift fur Angewandte Physik, Band 6, Heft 11, 1954, S. 499 - 507; Gerhard Piefke "Die Ausbreitung Elektromagnetischer Wellen in einem PYRAMIDENTRICHTER".

I was also invited by the Patent Office to tell something about the amount of microwave energy coming from the Sun, eventually reflected by the Earth, with regard to the possibility of resonant action on the micro-structure grid of the razor blade edge, and I proved by scientific evidence that with help of the pyramidal resonant cavity or by the bunching effect of the pyramid as a resonant horn that this energy would be sufficient.

PYRAMID POWER II

I proved that the energy needed for the steel-crystal grid action on dislocations is only 1 to 1.5 eV (electron volts - one electronvolt equals or represents the energy of 1.6 x 10 w/sec) this means that the energy is so low that it can easily be obtained from 'sferics' and also from 'technics' (microwaves produced by technical devices in the surrounding environment). See for example: P. Fischer und Kochendorfer "Plastische Eigenschaften von Kristallen (Kristallgittern) und Metallischen Werkstoffen" (Plastic Characters of Steel Crystal Grids).

My hypothesis, elaborated for the Patent Office (I do not affirm here that it is the only one possible) explains also, why the regenerating pyramid should not be placed too near the walls of the room or near large metal masses, or near electrical devices (surely not a television set).

If the 'man on the street' asked me how this little and simple device can act on a razor blade edge, my popular explanation compares the pyramid with a photo-cell, which acts without an artificial source of energy, using only visible Sun-light, whereas my device acts with an invisible Sun-light!

This is the main part of my hypothesis for the Patent Office, which at the end of the examination period, after 'only' 10 years, with the final proof of real functioning, given by the Chief Examiner, was compelled to allow the release of the patent for issuance.

From all this, we see that no 'magic' is needed to explain the function of the razor blade pyramid, and at the same time also the mummification effect of the model pyramid. There are two main factors:

> (1) Fast Dehydration: As I explained, also possible in some manner in the razor blade crystal edge.

> (2) Action on the microcrystalline lattice of inorganic matter (fine layer of steel), or action on the strucutre or microstructure of organic matter, living or dead, sterilization (killing of micro-organisms). It must be emphasized, that this action can in extreme cases also kill small animals by the rapid dehydration and some 'devitalization'.

I have spoken about other models than the Cheops geometry. The elevation of the Cheops walls is cca. 3 times 51, that means 51 degrees 51 minutes and 51 seconds (Piazzi Smythe, England; Abbe Moreux, France; L. Seidler, USSR). I, and also some French experimenters have found that a very good functioning pyramid model is also with a 65 degree elevation (what in Europe is cca. the magnetic inclination angle). I named this type the 'Inclination Pyramids'.

Upright to the wall of this model we found an elevation angle of 25 degrees, and this form represents a very good mummification pyramid with a great surface area on the walls, I call this the 'Contra-inclination Pyramid'. With all these models, I have made a great number of mummifications, but for the razor blade I preferred the Cheops model.

In the #9 1973 issue of the Journal 'Esotera' (BRD) on page 799-800 Hans Joachim Hohn confirms the functioning of the 'Cheops' on razor blades but he proposes his own model with an elevation angle of 69 degrees 20 minutes, baseline 15 cm, height of 22.2 cm, with which, as he says he has shaved very good 196 times, using a Wilkinson-sword blade. The title of his article is "In der Pyramide wird jede Klinge Wieder Scharf" (Every razor blade recovers its sharpness in the Pyramid Model).

How much easier was this 'pyramid job' for Mr Antoine Bovis, the indirect initiator of my experiments with the cardboard models. For him his intuition instead of scientific evidence was fully sufficient. With his divining rod or his pendulum of radiesthesia he measured all imaginable and also unimaginable, and it was probably with the pendulum, that he found the possibility of mummification in little Cheops models.

He used for this a cardboard model with approximately the same proportions of the Great Pyramid (where traveling in Egypt, he found the mummified remains of animals in the King's Chamber, 1/3 above the base of the pyramid, what was probably for him 'intuition shock'), but in a scale of 1:1000 (15 cm high) and 1:500 (30 cm high), the baseline was calculated by multiplying the height with the 'magic number pi/2' cca. 1.57.

He was sure that his 'mad seeming' experiments of mummification would succeed without the help of any technical literature, Physical Revues, etc. For him it was fully sufficient to use his little pendulum of patented construction (in France it is possible to obtain a patent for all things, sensible or nonsensible, without any proof, it only has to be new. It is not necessary to prove that it works! How easy for inventors of all kinds).

Intuition is at the base of nearly every great invention in physics; we can recall a great many glorious names of scientists such as Planck, Einstein, Mendelejev, Kekule, etc., all used intuition very often without being conscious of it!

I found the name of A. Bovis in a little booklet on radiesthesia which had several of his lectures to the radiesthesia circles of Nizza. This booklet was about his numerous inventions for measuring polarity of all imaginable things (also the numbers!) with his little special 'magnetic pendulum Bovis', the best of all, as he says. Following each paragraph, there was another "Law of Radiesthesia", the only ones possible, all found by Bovis!

In one of his lectures he spoke about mummification experiences with the Cheops Pyramid cardboard model, because he found in this model (using his pendulum) the 'same radiations' as in the King's Chamber of the Great Pyramid. One thing is evident: his models worked! He spoke about mummification of dead organic matter, meat, eggs, also of little dead animals, with the same result as he found in the Great Pyramid at Giza, where cats, dogs, and other animals were found to be perfectly mummified.

Because it was relatively simple to ascertain, if it is all only fiction or reality, I constructed a great model Cheops 30 cm high, in cardboard 3 mm thick (scale 1:500) and to my greatest astonishment I was also, as Mr Bovis, able, to make mummifications, to repeat with success his mummification experiments, mummifying beef, calf or mutton meat, eggs, but also flowers and little dead animals such as frogs, snakes, a lizzard, etc. I informed Mr Bovis about my experiments and had some agreeable correspondence with him, with reserve, that he was a little 'too magic' for me, a radio engineer. He found with his pendulum, radiations in all that he touched.

By his letters, I came to know that Bovis had an ironmongery (hardware: ed) shop at Nizza ('Quincaillerie Bovis & Passeron') and that he was considering himself as a great inventor in radiesthesia laws and also of devices of all kinds, he had another firm: 'Artisanat A. Bovis, Nice', for the fabrication of radiesthesia devices.

Some of his manufactured products were: Pendulum 'paradiamagnetic' (as he called it), 'radioscope, biometer, Plates **Magnetics** for mummification and action on liquids, material: brass or copper, nonmagnetic, something similar to the Flanagan Plates but constructed and sold already from 1931.

I made a great number of mummifications with pyramids of different forms and shapes, but most of them were with the Cheops-type, and I published something about it in collaboration with Mr Martial from Valenciennes, in French and Belgian radiesthetic journals, e.g: 'La Revue Internationale de Radiesthesie' # 7, Avril 1948, pg 54 -57 (France), or 'La Radiesthesie pour Tous" # 12 1949, pg 377-379 (Belgium) and have contacted herewith also other French radiesthesiests interested in mummification by Cheops pyramid models.

PYRAMID POWER II

In the journals of radiesthesia I found other experimenters as Mr Lombart with his 'Contra-inclination Pyramid' of very good functioning as he says, further Mr Pierre Bories with precisely conducted experiments, and others. As a radio engineer I was forced to admit that there is something very strange in this phenomenon, because some energy must be evidently concentrated in the pyramid model.

It was just the 'looking for the nature of this energy' which had encouraged me to make a further 'mad experiment': to give a new razor blade of good quality (the Blue Gillette) in the cardboard Cheops pyramid and see, if it should not become blunted. I was looking for some physical evidence of a concentrated force, acting in the pyramid.

So began my razor blade adventure with the Cheops model. My supposition, that the blade in the pyramid would lose its sharpness was false. Just the contrary occured, and when I shaved 50 times instead of the usual 8 times, I was forced to admit that my supposition was wrong indeed.

The first razor blade experiment was made with a Cheops-type pyramid, 15 cm high (scale 1:1000), the blade was posed horizontally North-South with its longitudinal axis and at a level of 1/3 the height from the base, the two sides of the

PYRAMID POWER II

pyramid oriented in the same longitudinal direction.

By numerous experiments I found that for this purpose a pyramid of 8 cm high in cardboard or 7 cm high in styrofoam is sufficient. Years later this styrofoam model was fabricated by a plastic material factory 'Plastimat', but only some hundreds of pieces, perhaps because a great factory of razor blades, made in foreign license had no interest in using blades 100 times or more. I say 'perhaps' as I do not know all the circumstances.

But it is for everyone so easy to make for himself this little pyramid in cardboard with the help of some instructions in journals about this invention, e.g. the article of Mr Z. Rejdak in the Czech journal 'Signal' # 34, August 8, 1968. It is difficult for me to guess, how many home made pyramids exist in the USSR. Not one of the thousands and thousands of users has written to me with any complaints, but a great number have written with great enthusiasm.

I can say, that the last 25 years are for me a long experimental sequence and each shaving an experimental evidence, which can sometimes inform me by an unexpected change in sharpness about a change in some meteorological or cosmic disturbance (e.g. of the Sun). The edge in the pyramid is a 'living entity', it is in connection with the environmental field, and the greatest surprise is perhaps a bad shaving day is followed immediately by an excellent shaving day. A temporary weakness of the edge doesn't mean that the blade is a reject. Try one and see for yourself.

To judge the sharpness of the blade I have introduced (the same as the great Czech factory 'Astra' in their examination of new blades) a scale with six degrees: No 6: Excellent, No 5: Very Good. No 4: Good, No 3: Sufficient and No 2: Baddish, No 1: Insufficient. In the first five years and three months of my blade-pyramid experiments, from March 3, 1949 to July 6th, 1954 the average value of one blade was 105 shavings (using in this period only 18 blades of different makes) and as a maximum there were numbers like 200, 170, 165, 111, 100 and so on, with each blade. In 25 years, I have used only 68 blades!

My correspondence about his strange patent is not only with a great number of states in Europe, but also with the USA, South America, Australia, and New Zealand. The USSR also showed great interest, where e.g., in the 'Komsomolskaja Pravda' from October 10, 1970 Mr Malinov, C. Sc., wrote an interesting article about a 'strange invention' - as he said, as a physicist formed a logical explanation of the functioning by using electromagnetic theory combined with the Earth's magnetic field, and also with the Lorentz forces. My little pyramid, home-made, is in customary use in the USSR. The same article, by Mr Malinov

PYRAMID POWER II

was published in 1973 in the Journal 'Heureka' of Moscow.

It seems to me, that I can already finish this long description of a very fancy invention. I must add, that I wrote about the regenerating pyramid some articles in popular science journals in Czechoslavakia and other Slavik countries. I have also done a great many television shows about it which have all brought me a great number of friendly letters confirming or demanding more information.

The interest in my invention came, when the patent had expired, in 1967. In the end, I wish that everyone who uses the pyramid will obtain 200 or more shaves from one blade!

      KAREL DRBAL
      FEBRUARY, 1975

PYRAMID POWER II

KAREL DRBAL
REGARDING PYRAMID RESONANCE
CALCULATING ELECTROMAGNETIC WAVELENGTH

From a long series of pyramid shaped hats constructed in cardboard I have introduced one new designation (this experiment was made twenty years ago) the Pyramid Resonance Wavelength (PRWL) or shorter 'L' (Lambda) which I calculated as a microwave corner horn pyramid antenna with a quadrangular base, is easy to calculate. If we designate the height of the pyramid as 'v', the altitude of the pyramid face triangle as 'h', and the resonance wavelength as 'L', the resonant wavelength is then $L = 2(h - v)$. See Radio Technical Journal RADIO TECHNIK, Zeitschrift fur Hochfrequenztechnik, Heft 11, November 1948 "Dr Techn. W. Nowotny: Neuere Bundelungsprinzipien fur Mikrowellen, Hornstrahler, Linsenantennen, Schlitzantennen, S. 570, Abb. 2". E.g. a Cheops shaped pyramid 15 cm high, 23.6 cm base, 19.1 cm height of the triangle, has the resonance wavelength $L = 2(19.1 - 15.0) = 8.2$ cm.

THE GREAT PYRAMID - ITS DESIGN CONCEPT
E. D. ROBINSON

THE GREAT PYRAMID - ITS DESIGN CONCEPT

E. D. Robinson

INTRODUCTION

Only three design concepts for the Great Pyramid have come into being since its completion. One was an unreasonable concept described by the Greek historian, Herodotus, in the 5th century B.C. On p.124 of <u>Historie</u>, Book II, Herodotus wrote that the pyramid was square with sides of 8 plethra and that the height was the same. He said that he measured it. Eight plethra translates to 246.25 meters and the base as we know it is just over 230 meters. Given the dimensions by Herodotus, the face-to-base angle would be greater than 63 degrees, or an angle slightly less than the angle of the pyramid on the Great Seal of the United States.

The second one, called the design by PI concept, was developed by an English publisher, John Taylor, in the mid-1850's. Taylor used dimensions from surveys of the Great Pyramid in 1799-1801 and in 1837. He divided the perimeter of the pyramid (using a basewidth of 764 feet) by twice its vertical height (using a height of 486 feet) and obtained a number, 3.144033, which he recognized as being very close to the value of PI, 3.1415927. He theorized that the ancient architects of the Great Pyramid knew the value of PI to three digits and designed the pyramid accordingly. Taylor's theory was picked up and supported by the first proclaimed pyramidologist, Charles Piazzi Smyth.

The third design concept came into being by some unknown author sometime prior to 1921. The theory claims that the slant height of the pyramid is proportional to the Golden Number as one-half the base-width is proportional to unity. The vertical height, then, is proportional to the square root of the Golden Number. The Golden Number is commonly denoted by the Greek letter PHI (ϕ). The design philosophy is referred to as the design-by-ϕ concept.

Two highly accurate surveys of the Great Pyramid have been made since Taylor's time: one by W.M.F. Petrie in 1881-82 and another by J.H. Cole in 1925. Petrie's survey included the internal features as well as the main body while Cole's was devoted to determining the exact size of the main body and its orientation relative to true north.

The results of the 1881-82 and 1925 surveys will be used to perform an error analysis to the design-by-pi and the design-by-phi concepts.

THE SURVEY DATA

It is doubtful that a modern survey of the Great Pyramid would provide data which is more accurate than that obtained by W.M.F. Petrie in 1881-82[1] and J.H. Cole in 1925[2]. Petrie determined that the earlier surveyors[3] had uncovered the sockets at the northeast and northwest corners of the pyramid which were assumed to hold the outer casing stones at bedrock level. He found that these two sockets were at different elevations. He selected the upper surface of the pavement around the north base of the pyramid as a reference to make measurements in the vertical plane. He found the northeast socket to be 28.5 inches below the level of the pavement and the northwest socket to be 32.8 inches below the level of the pavement. Earlier surveyors had erected verticals to the outer corners of the sockets and measured the distance between verticals. This, then, was the dimension they gave as the basewidth of the Great Pyramid and it was the basewidth used by Taylor.

Petrie theorized that the locations of the sockets at different elevations came about because of a desire of the builders to construct the pyramid on unleveled bedrock with minimum effort. He assumed that the intended basewidth was that in the level plane of the upper surface around the pyramid. His average basewidth in the plane was 9068.8 ± 0.5 inches ($230,347.5 \pm 12.7$ millimeters).

Cole used the same plane of reference and obtained an average basewidth of 230,363.75 ± 6 to ± 30 millimeters (9069.439 ± 0.24 to ± 1.18 inches). Cole's survey is generally accepted to be the more accurate and his average basewidth will be used in the error analysis to follow.

Petrie made eight sets of angle measurements at the north face of the pyramid. For each set, he listed the mean angle and range of uncertainty about it. His data is given in Table A. He used three measurement techniques with no one technique being superior to the other. No one set of data appears to be more accurate than another set, therefore the data in the eight sets must be used in any error analysis.

TABLE A - PETRIE'S ANGLE DATA

SET	DATA SOURCE	MEAN VALUE	RANGE OF UNCERTAINTY
1	Casing stones, in situ, N Base, by theodolite to three points on top and three on base.	51°46'45"	± 02'07"
2	Same, by Goniometer and Level	51°49'	± 01'
3	Same by square and level	51°44'11'	± 00' 23"
4	Same with 5 points overthrown by Goniometer	51°52"	± 02'
5	18 Fragments, by Goniometer to 3 points top and bottom	51°53"	± 04'
6	N face, by entrance, passage	51°53'20"	± 01'
7	N face by air channel mouth	51°51'30"	± 00' 20"
8	N face, existing surface, 48 points, top and bottom by theodolite	51°50'40"	± 01' 05"
	AVERAGE MEAN ANGLE	51°50'03.25"	

Given the mean angles and the range of uncertainty about it, the data can be handled in several ways to establish an average angle and a range of uncertainty about it. The variance for each set of data can be determined and the standard deviation (one sigma value) for the eight sets of data can be calculated. The results are shown in the last column of TABLE B. The average angle is 51°50'03.25" and the root-mean-square deviation about the average angle is 03' 0.233".

A second way of evaluating the data is to accumulate the minimum and maximum angle using the range of uncertainty for each set of data. The average minimum and maximum angles are then determined. The results are shown in TABLE B and this indicates an average angle of 51°50'03.25" with an uncertainty of ± 01' 29.375" about it. The second method provides a 'window' about the average angle which is about one-half as wide as the window provided by the first method. The narrowest window will be used in the analysis.

TABLE B - PETRIE'S ANGLE DATA

CASE	MINIMUM ANGLE	MEAN ANGLE	MAXIMUM ANGLE	$(B_x - B)^2$
1	51.743 888 89°	51.779 166 67°	41.814 444 44°	3.032 645 $\times 10^{-3} \text{deg}^2$
2	51.800	51.816 666 67	51.833 333 33	0.308 686 "
3	51.730	51.736 388 89	51.742 777 78	9.574 082 "
4	51.833 333 33	51.866 666 67	51.900	1.051 740 "
5	51.816 666 67	51.883 333 33	51.950	2.410 535 "
6	51.872 222 22	51.888 888 89	51.905 555 56	2.986 924 "
7	51.852 777 78	51.858 333 33	51.863 888 89	0.580 675 "
8	51.826 388 89	51.844 444 44	51.862 500	0.104 210 "
B	51.809 409 73°	51.834 236 13°	51.859 062 51°	
D^2				2.506 188 3 $\times 10^{-3} \text{ deg}^2$
D_{rms}				0.050 062 degrees

B = average angle

B_m = average of mean angles

B_x = case mean angle

E.D. ROBINSON - PYRAMID DESIGN CONCEPT

ANGLE ANALYSIS

Taylor used a basewidth of 764 feet and a vetical height of 486 feet. This defines a right triangle with a base angle of 51°49'56.5" which is only 6.75 minutes of arc less than the average mean data from the survey.

C. Piazzi Smyth, the Astronomer Royal of Scotland in the 1860's, picked up on Taylor's theories and spent over four years trying to prove them.[4] Smyth assumed an angle of 51°51'14.3" (pg 26 of ref. 4) which he incorrectly stated to be the mean between the angle measured by Howard-Vyse and one calculated by John Hershel[5]. The angle assumed by Smyth is 1' 11.05" greater than the average mean angle from the survey data and it is 20.575 minutes of arc inside the upper limit of the narrowest range of uncertainty.

André Pochan[6] pointed out that if the design-by-ϕ philosophy had been used, some unique trigonometric relationships would have occurred. One relationship is:

Angle of face = arccos ϕ^{-1} = arctan $\phi^{1/2}$ = 51°49'38.25" where ϕ = $1.25^{1/2}$ + 0.5, the golden number. The angle of the design-by-ϕ philosophy is 25 minutes of arc less than the average mean angle from the survey data.

E.D. ROBINSON - PYRAMID DESIGN CONCEPT

In the design-by-pi philosophy, the face angle is:

face angle = arctan 4/PI = 51°51'14.3065".

The face angle is 1' 17.8" greater than the angle from the survey data and it is 18.32 minutes of arc inside the upper limit of uncertainty.

The angles can be summarized as follows:

upper limit of uncertainty = B_{max} = 51.859 062 51°

by the PI philosophy = arctan 4/PI = 51.853 974 02°

mean survey data = av. mean = 51.834 236 13°

by Taylor's dimensions = arctan 486/382 = 51.832 361 68°

by the ϕ philosophy = arctan $\phi^{1/2}$ = 51.827 292 39°

lower limit of uncertainty = B_{min} = 51.809 409 73°

It is of interest to note that the design concepts give angles which are inside the limits of uncertainties and that the angle using Taylor's dimensions is the closest one to the mean survey data.

E.D. ROBINSON - PYRAMID DESIGN CONCEPT

The evaluation of the angle, therefore, fails to produce conclusive evidence for deciding on the design philosophy which the ancient Egyptians had in mind.

TAYLOR'S THEORY

Taylor divided the perimeter of the pyramid (his 3065 feet) by twice the vertical height (his 972 feet) and obtained a number, 3.14403..., which is

0.0777% greater than the value of PI

In the design-by-ϕ concept, the basewidth is proportional to two as the vertical height is to the square root of the golden number. The perimeter, in this design philosophy, divided by twice the vertical height is:

$$4/\phi^{1/2} = 3.144605$$

and this is 0.0959% greater than the value of pi, however it is only 0.018% greater than Taylor's number. Taylor's finding is much closer to $4/\phi^{1/2}$ than it is to the value of PI.

A search of Taylor's works and those of Smyth who followed him fails to reveal that either of them had a knowledge of the golden number or functions of it.

E.D. ROBINSON - PYRAMID DESIGN CONCEPT

Since Taylor was a mathematician and an amateur astronomer, he most certainly was familiar with the value of PI and its uses. He may have heard of the golden number and the golden ratio, however he probably had no reason, ever, to have divided four by the square root of the golden number. Therefore, he failed to recognize the closeness of his number to a fraction of the golden number.

Since Smyth used Taylor's works extensively (those on linear measures as well as those on the Great Pyramid), had corresponded with him, and had visited him, it appears that a mention of the golden number or golden ration by Taylor would not have gone unnoticed by Smyth. If Taylor had mentioned it, Smyth most certainly would have picked it up and looked into it. In the voluminous Our Inheritance in the Great Pyramid, Smyth developed some two dozen theories and attempted to prove them by the selection of data which would best fit his purpose. His <u>assumption</u> of an angle of $51°51'14.3"$, seemingly proved his and Taylor's theory that the ratio of the pyramid's vertical to twice the basewidth is the same as the ratio of unity to PI is unforgiveable. Heated debate over this and other theories by Smyth resulted in his resignation from the Royal Society.

A search of the works available to both Taylor and Smyth, such as those by Greaves, Jomard, and Howard-Vyse[3], fails to reveal a consideration of the design-by-φ philosophy. If the philosophy had been studied, Taylor and Smyth certainly would have picked it up. The first to refer to the design philosophy appears to be Borchardt in 1921 who erroneously quoted Herodotus as writing, "the design is such that the area of the face is the same as the square of its (vertical) height".

It can be argued that Taylor failed to consider the design-by-φ concept because of a lack of familiarity with the golden number and functions of it. He most likely was led to the design-by-PI concept because of his familiarity with PI and its use. His overwhelming enthusiasm following the connection stimulated Smyth to set out to gather data which would support that and other theories. Smyth was unscrupable in using data to back a theory. An Egyptian Inspector of Antiquities came upon Smyth and an aid filing the outer surfaces of a boss stone in the Antechamber to make it a 'pyramid inch' high.

There is no evidence in or around the Great Pyramid to lead one to believe that the ancient Egyptians who designed and built the structures had a need to know the value of PI. Did they have a need to know the golden number? The following analysis will provide an affirmative answer to the question.

DIMENSIONS OF THE GREAT PYRAMID

One of the problems of evaluating design concepts is the availability of accurate data. TABLE B provides eight sets of angle data: the mean angle for each set and a range of uncertasinty about it. It is generally accepted that Cole's average basewidth is more accurate than Petrie's. Using Cole's average basewidth, the vertical and slant heights can be determined for each set of data. The results are shown in TABLES C and D. The tables show the minimum, mean, and maximum dimensions for each set of data. The standard deviation and the average minimum and maximums are given in the tables. The dimensions can be invaluable to those who read the literature on the Great Pyramid and try to sort out truths from the many claims which have been made and which cannot be supported by the evidence.

The golden number is 1.618 033 988 7.. ..., an irrational number. Suppose that the averages in TABLE C are divided by one-half the basewidth (377.893 290 7 feet) and the result squared.

TABLE C - VERTICAL HEIGHT FROM SURVEY DATA

CASE	MINIMUM, FEET	MEAN, FEET	MAXIMUM, FEET	VARIANCE, FT.^{2}
1	479.250 207 6	479.857 581 9	480.465 906 6	0.904 376 274 2
2	480.216 713 9	480.504 258 2	480.792 015 6	0.092 605 002 1
3	479.011 343 9	479.121 203 1	479.231 093 0	2.847 203 343 3
4	480.792 015 6	481.368 169 9	481.945 178 5	0.313 153 279 2
5	480.504 258 2	481.656 567 1	482.812 298 9	0.719 100 947 2
6	481.464 278 3	481.752 747 3	482.041 429 8	0.891 472 851 0
7	481.128 007 1	481.224 051 2	481.320 124 1	0.172 625 541 6
8	480.672 090 8	480.983 972 2	481.296 103 7	0.030 766 317 6
AVER.	480.379 863 9	480.808 568 9	481.238 019 0	
D^2				0.746 412 944 4 FT^2
D				0.863 951 934 1 FT

TABLE D - SLANT HEIGHT FROM SURVEY DATA

CASE	MINIMUM, FEET	MEAN, FEET	MAXIMUM, FEET	VARIANCE, FT.2
1	610.314 755 5	610.791 812 4	611.269 847 6	0.558 968 859 2
2	611.073 998 4	611.299 992 9	611.526 206 7	0.057 341 905 7
3	610.127 205 4	610.213 459 8	610.299 745 8	0.758 262 210 0
4	611.526 206 7	611.979 292 3	612.433 257 0	0.193 457 202 3
5	611.299 992 9	612.206 164 5	613.115 857 7	0.444 502 090 8
6	612.054 891 7	612.281 837 6	612.502 003 4	0.551 132 518 7
7	611.790 399 7	611.865 938 5	611.941 501 4	0.106 591 737 0
8	611.431 924 3	611.677 137 6	611.299 608 3	0.018 956 608 4
AVER.	611.202 421 9	611.539 454 4	611.877 253 6	
D^2				0.461 151 641 5 FT^2
D				0.679 081 469 0 FT

FIND THE FOLLOWING

from av. mins. - 1.615 962 31 (0.128%<φ)

from av. means - 1.618 847 859 (0.05%>φ)

from av. maxs. - 1.621 741 005 (0.229%>φ)

The average means divided by one-half the basewidth and the quantity squared is very close to the golden number (5 parts in 1,000 parts).

    Suppose that the averages in TABLE D are divided by one-half the basewidth. Find the following:

from av. mins. - 1.617 393 214 (0.039%<φ)

from av. means - 1.618 286 086 (0.0156%>φ)

froms av. maxs - 1.619 179 987 (0.0708%>φ)

    The ideal vertical height would be 480.687 691 4 feet which is only 1.45 inches shorter than the vertical height indicated by the average of the means from the survey data. The ideal slant height would be 611.444 188 6 feet and this is only 1.1432 inches less than that indicated by the survey data.

    The above indicates two things: one is that the ancient Egyptians knew the golden number and the square root of it; and they could cut and set huge blocks with machinists accuracy. What else did they know? The following indicates that they knew quadratics and a system of logarithms some 4200 years before Napier developed logarithms.

ANALYSIS OF AREAS

Suppose that the average mean dimensions from the survey data are used to find the areas of the facets of the pentahedron represented by the Great Pyramid (four triangular faces and a square base). The ideal dimensions can be found from the equations in TABLE E. TABLE F gives some ideal dimensions and dimensions from the survey data for comparison. The student can analyze the data -- the differences are very small.

Similar dimensions generated by the design-by-PI concept are not given, however they are very near the upper limit of uncertainties while the design-by-ϕ dimensions are very close to the average of the means.

The data supports an argument that the design of the Great Pyramid is such that the area of its very nearly square base times the square of the golden number is the area of the ideal pentahedron. What is the meaning of these dimensions mathematically?

E.D. ROBINSON - PYRAMID DESIGN CONCEPT

TABLE E - DESIGN-BY-ϕ PARAMETERS

PARAMETER	DIMENSION
Basewidth, av.	$Bw = 755.786\ 581\ 4$ ft
Vertical Ht	$\frac{Bw}{2} \phi^{1/2}$
Golden No., ϕ	$\phi = 1.25^{1/2} + 0.5$
Slant Ht	$\frac{Bw\phi}{2}$
Area, triangular face	$\frac{Bw^2 \phi}{4}$
Area, 4 faces	$Bw^2 \phi$
Area, pentahedron	$Bw^2 \phi^2$
Volume	$\frac{Bw^3}{3} \phi^{1/2}$
Angle, face plane to base plane	$\arccos \phi^{-1}$
Angle, apex of pyramid	$2 \arcsin \phi^{-1}$
Angle, face at the base	$\arctan \phi$
Angle, face at apex	$2 \operatorname{arccot} \phi$

TABLE F - COMPARISON OF IDEAL DIMENSIONS WITH THOSE FROM SURVEY DATA

DIMENSION	IDEAL DIMENSION	FROM SURVEY DATA*
Basewidth	755.786 581 4 feet	SAME
Area of Base	571,213.3566 feet^2	SAME
Vertical Height		
Average Minimum	- feet	480.3799 feet
Average Mean	480.6877 feet	480.8086 feet
Average Maximum	-	481.2380 feet
Slant Height		
Average Minimum	-	611.2024 feet
Average Mean	611.4442 feet	611.5395 feet
Average Maximum	-	611.8773 feet
Area of Four Faces		
Average Minimum	-	923,877.1782 feet^2
Average Mean	924,242.6259 feet^2	924,386.6276 feet^2
Average Maximum	-	924,897.2356 feet^2
Area of Pentahedron		
Average Minimum	-	1,495,090.535 feet^2
Average Mean	1,495,455.983 feet^2	1,495,599.984 feet^2
Average Maximum	-	1,496,110.592 feet^2
Slope of Face	51° 49' 38.2525"	51° 50' 03.25"

*Petrie's angle data and Cole's average basewidth.

E.D. ROBINSON - PYRAMID DESIGN CONCEPT

THE GREAT PYRAMID AND QUADRATICS

The areas in the pentahedron represented by the Great Pyramid can be described as follows. Let the base be one unit square (one unit equals 755.786 581 4 feet). Then the area of the four triangular faces is ϕ square units, and the area of the pentahedron is ϕ^2 square units. This generates the following equation:

area of pentahedron - area of 4 faces - area of base = 0

or, $\quad \phi^2 u^2 - \phi u^2 - 1 u^2 = 0$

The above equation can be written as:

$$\phi^2 + b\phi + c = 0$$

when $b = c = -1$. The solution by completing the squares yields two roots: the golden number and the negative of its reciprocal. The relationships of the areas in the pentahedron can be described by the 'pure' quadratic equation which yields, as one root, the golden number.

THE GREAT PYRAMID AND LOGARITHMS

It can be shown that $\phi^0 = 1$. Therefore the three functions above (ϕ^2, ϕ, and ϕ^0) are three functions in an irrational number series which is both logarithmic and arithmetic in nature. That is, a function multiplied by the golden number is the next larger function in the series and conversely, a function divided by the golden number is the next smaller function in the series. A function in the series added to the next larger function is the second larger function in the series, and, conversely, the difference between two consecutive functions is the next smaller function in the series. The golden series is logarithmic to the base ϕ.

CONCLUSIONS

It has been widely accepted as fact that the design of the Great Pyramid incorporates the value of PI. This theory was advanced by John Taylor in 1859[7] and supported by C. Piazzi Smyth in 1877 who used an <u>assumed</u> angle to do so. An unknown author <u>suggested</u> a design-by-ϕ concept sometime prior to 1921 (Borchardt's time) and André Pochan partially supported it. The analysis herein has established a highly correlative connection between survey dimensions and the design-by-ϕ philosophy.

If the ancient architects and builders of the Great Pyramid used the design-by-ϕ philosophy, then they had a knowledge of geometry and mathematics which few historians will give them credit for.

Elmer D. Robinson

December 7, 1978
Wheaton, Maryland

REFERENCES

1. W.M.F. Petrie, THE PYRAMIDS AND TEMPLES OF GIZEH, Field and Tauer, London; and Scribner and Welford, New York 1883; revised 1885.

2. J.H. Cole, DETERMINATION OF THE EXACT SIZE AND ORIENTATION OF THE GREAT PYRAMID, Survey of Egypt, Paper #38, Government Press, Cairo 1925.

3. John Greaves, PYRAMIDOGRAPHIA: OR A DESCRIPTION OF THE PYRAMIDS OF EGYPT, Geo. Badger, London 1646; and J. Brindley, London 1740.
-- Edmè Francois Jomard and Jean Marie Joseph Coutelle, DESCRIPTIONS DE L'EGYPTE, French Government, Twenty-one Volumes, Paris 1809 to 1822.
-- Richard Howard-Vyse, OPERATIONS CARRIED ON AT THE PYRAMIDS OF GIZEH IN 1837-38, J. Fraser, Three Volumes, London 1840 to 1842.

4. C. Piazzi Smyth, OUR INHERITANCE IN THE GREAT PYRAMID, Dalby Isbister & Co., London 1877; and Rudolph Steiner Publications, New York 1977.
-- C. Piazzi Smyth, LIFE AND WORK AT THE GREAT PYRAMID, Edmonton & Douglas, Edinburgh 1865.

5.   John Hershel, RS, POPULAR LECTURES ON SCIENTIFIC SUBJECTS, reprint of ATHENIUM, April 23, 1860, W.H. Allen, London 1880.

6.   Andre Pochan, L'ENIGME DE LA GRANDE PYRAMIDE, Editions Robert Laffant, Paris 1949; and THE MYSTERIES OF THE GREAT PYRAMIDS, Avon Books, New York 1978.

7.   John Taylor, THE GREAT PYRAMID: WHY WAS IT BUILT AND WHO BUILT IT?, Longmans, London 1859.

PYRAMID POWER II - MODEL

THE GEOMETRY OF THE GREAT PYRAMID
E.D. Robinson

INTRODUCTION
The Great Pyramid of Giza is the most spectacular of all pyramids. It is the largest and most geometrically perfect. There is little doubt that the ancient architects of the Great Pyramid had a knowledge of geometry and mathematics which few historians will give them credit for. There is little doubt that the builders of the Great Pyramid were highly skilled stone masons who could shape the massive stone blocks with machinists precision. They also must have been versed in astronomy and geodetics since the Great Pyramid is oriented almost perfectly to True North. If they were so skilled, it is reasonable to assume that the ancient architects designed the Great Pyramid on their equivalent of our modern drafting tables and constructed accurate scale models of the Great Pyramid.

CLUES TO A PLANE GEOMETRICAL MODEL
An analysis of the Great Pyramid's geometry, both in its main body and in the internal constructions, using highly accurate data from W.M.F. Petrie's 1881-82 survey indicates that the slope of the decending tunnel is almost 1-to-2. That is, for each two units in the horizontal plane the tunnel rises or drops one unit in the vertical plane. The slope is given by a triangle having a base of two units, an altitude of one unit and a hypotenuse of the square root of 5 units ($5^{1/2}$). The

PYRAMID POWER II - MODEL

sloping angle is the arctangent of 1/2. The slope of the ascending tunnel is slightly less than this and may have been sloped less by intent.

The analysis indicates that the sloping height of a face of the Great Pyramid is proportional to the golden number, represented by the Greek Letter ϕ, as one-half the basewidth is proportional to unity. The apex angle of a face is twice the arctangent of the reciprocal of the golden number. The golden number is one-half plus the square root of 1.25.

If the angle of the descending tunnel is added to the apex angle of a face, one obtains an angle of 90 degrees. Four of these combinations make a circle. This suggests that a model of the Great Pyramid can be laid out on a plane surface and the geometrical construction folded to produce an exact scale model of the Great Pyramid.

THE GEOMETRICAL CONSTRUCTION

Refer to the following figure to visualize the following steps in the geometrical construction.

· Draw a horizontal line AB with a length which is 1.7013 times as long as the basewidth of the model to be constructed. The equation for the multiplication factor is: AC = desired basewidth $\times \frac{(\phi+2)^{1/2}}{(1.25)^{1/2}}$

where $\phi = 0.5 + (1.25)^{1/2}$

PYRAMID POWER II - MODEL

PLANE CONSTRUCTION OF PYRAMID MODEL
©1978 E.D. Robinson

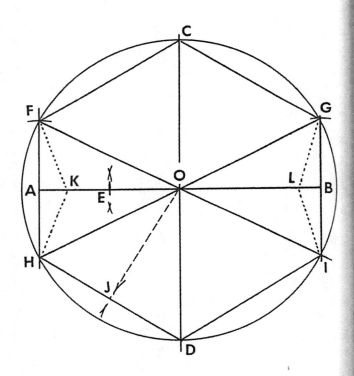

PYRAMID POWER II - MODEL

• Bisect AC to find point O. Draw a vertical CD, through O.

• Bisect AO to find point E.

• Draw a vertical through points A and B.

• Measure AE with a compass and swing the compass about A to find point F on the vertical through A.

• Swing the compass about B to find a point G on the vertical through B.

• Draw a straight line through points F and O and find point I on the vertical through B.

• Draw a straight line through points G and O and find point H on the vertical through A.

• The slope of the lines is the ideal slopes of the ascending and descending tunnels. FO represents an edge of the pyramid.

• Measure FO with a compass. With center at O, scribe a circle F C G I D H.

PYRAMID POWER II - MODEL

• Draw lines from F to C, C to G, I to D, and D to H. The isoceles triangles FOC, COG, IOD, and DOH represent the four sides of the Great Pyramid with their apex at point O.

• Bisect line DH to find point J. Connect J and O. JO represents the slant height of the Great Pyramid.

SHAPING THE MODEL
The layout should be made on material such as construction paper which can be folded and creased and which will be stiff enough to preserve flat surfaces.

Cut around the circle. Fold the construction paper up and around line AB and crease. Fold the construction down along lines FI, CD and GH creasing each fold. Fold the construction up along the bases FC, CG, ID and DH. The tabs will be used to glue the construction to a flat surface. Cut out two sections along the dotted lines to points K and L. Discard the material.

Hold the geometrical construction and tuck darts FOH and GOI downward and inward. Smear glue on the upper surfaces of the darts and bring FO to GH. Press the surfaces of the dart together. Bring GO to IO and press the surfaces of the dart together. Square up the model of the pyramid.

152

PYRAMID POWER II - MODEL

Smear glue on the underside of the four tabs. Mount the model on a plane surface. **It is an exact scale model of the Great Pyramid.**

RELATIONSHIPS IN THE MODEL

The vertical height of the model is proportional to the square root of the golden number. The slant height of a face divided by one-half the basewidth is the golden number. The area of the base times the golden number is the area of the four triangular faces. The pyramid is a pentahedron with a square base and four triangular faces. The area of the pentahedron is the area of the base times the square of the golden number.

The design of the Great Pyramid incorporates the basic quadratic equation:

$$X^2 + X + 1 = 0$$

This can be seen as follows: (area of pentahedron) - (area of four faces) - (area of base) = 0.

$$\phi^2 \text{ units}^2 - \phi \text{ units}^2 - 1 \text{ unit}^2 = 0$$

The above can be written as:

$$\phi^2 + b\phi + c = 0, \text{ where } b = c = -1$$

The solution to the above by completing the square yields two roots: one is the golden number and the other is the negative of its reciprocal.

PYRAMID POWER II - MODEL

CONCLUSIONS
Models of the Great Pyramid can be constructed with a square and a compass -- the primary tools of ancient Freemasons. Each element of the construction can be described mathematically and scaled to full size with mathematical precision. Let the model have a one-half basewidth of one unit. The following scale factors apply.

1 foot in the model = 377.893 290 7 feet in the pyramid

1 meter in the model = 115.181 875 0 meters in the pyramid

1 Memphis cubit in the model = 220 M. cubits in the pyramid

The much discussed face-to-base angle in the Great Pyramid is the arccosine of the reciprocal of the golden number. One half the apex angle of a face is the arctangent of the reciprocal of the golden number. Refer to the figure. It can be shown that the area of triangle FAO times the square root of five is the area of triangle FOC, and the square root of 5 is $2\phi - 1$. The golden number, ϕ, is prevalent throughout the design of the Great Pyramid, both in its main body and in its internal constructions.

This exercise raises another unanswered question: why did they choose the golden number and functions of it to design the Great Pyramid? Functions of the

PYRAMID POWER II - MODEL

golden series are logarithmic to the base ϕ. The golden spiral, an equiangular logarithmic spiral, is found in many places in nature. How much did they know that we still refuse to give them credit for?

        Prepared:
     August 12, 1978
      E.D. Robinson

PYRAMID POWER II - BIOGRAPHIES

BIOGRAPHIES

The scientists who performed the pyramid research described in this book are all top-ranked professionals in their respective fields and have published numerous papers.

At publication, we do not have all their biographies available, therefore we are printing short biographical sketches on the ones we do have.

PYRAMID POWER II - BIOGRAPHIES

THEODORE W. HORNER, PH.D.

As a senior statistician and applied management scientist, Dr. Horner formulates mathematical characterizations (mathematical models), develops procedures for testing hypotheses and constructing point and confidence interval estimates of parameters that apply uniquely to given data. This includes executing data collection plans, formulating and implementing the analysis of the resulting data, inclusive of the statistical design of experiments and surveys.

As a project leader in the Management Analysis Department of General Mills, he applied statistical techniques to operations research type problems with profit implications. His experience was further broadened by the application of statistics to defense research while at Booz Allen Applied Research, Inc., where he held the position of Principal Scientist for many years.

Dr. Horner received his Ph.D. degree in mathematical statistics from the Instutute of Statistics at North Carolina State University. Subsequently, he served as Assistant Professor and member of the graduate faculty of the Statistical Laboratory of Iowa State University. In this capacity he taught undergraduate and graduate students in both applied and theoretical statistics while at the same time serving as statistical consultant to the Iowa Agricultural Experiment Station, emphasizing the area of mathematical genetics. He also taught operations research at the Center for Engineering Studies at Vanderbilt University.

SANDOR HOLLY, Ph.D.
(Applied Physics Researcher)

Dr. Holly is currently Principal Research Scientist of the Research and Technology Department of an Alexandria, Virginia, based research corporation (since January of 1972). His primary responsibility has been to build up a strong laser-optics capability and to coordinate all optics and laser related activities for the company.

Prior to his present position, Dr. Holly was a Faculty member and Senior Research Associate with the Physics Department, University of Maryland (1970-1972). His task was mainly instrumentation oriented, aimed at improving the noise performance of the Lunar Surface Gravimeter (Apollo 17 Project) electronics package and analyzing the sources of noise. Other projects include: Analytical study of the radial resonant modes of thick metal disks; feasibility study of a cryogenic facsimile of a LaCoste-Romberg gravimeter sensor; and magnetic field sensitivity studies of the Lunar Surface Gravimeter.

PYRAMID POWER II - BIOGRAPHIES

While a full-time graduate student at Harvard University (1962-1968), he worked during the summers for Arthur D. Little Company, designing and operating a microwave apparatus for superconductive tunneling in thin film-microwave field interaction studies.

He received his Diploma in Physics and Mathematics (1955) from ELITE, University of Sciences in Budapest, Hungary. Subsequently, he attended for one year at the Technical University of Budapest and in 1958 completed all requirements for a B.S. degree in Electrical Engineering at M.I.T. in Cambridge, Massachusetts. He received his M.S. in Electrical Engineering (1960) from M.I.T., and M.S. degree in Physics from Harvard University (1962). In 1968, he received his Ph.D. in Applied Physics from Harvard University. Dr. Holly has subsequently taken several graduate courses in Modern Optics; Systems Engineering; High Power/High Energy Lasers; and Photoelectronic Imaging Devices.

PYRAMID POWER II - BIOGRAPHIES

JOHN E. STAUCH, Ph.D.
Clinical Laboratory Director

Dr. Stauch, a micro-bacteriologist and a successful business executive, has been in the profession since 1950 and has had extensive experience in directing a wide variety of Air Force and civilian medical laboratories. Among his current business and consulting activities are several laboratory directorships including the director of Physicians Bioanalytical Laboratory (Oxon Hill, Maryland); an Assistant Professorship in the Department of Medicine, George Washington, University; and he holds the rank of Scientist with the Armed Forces Institute of Pathology (Washington, D.C.).
He holds the distinction of having been appointed in 1967, to the Maryland Laboratory Advisory Council by then Governor Millard E. Tawes for a four year term.
Dr. Stauch received his B.S. degree in Pre-Medical (1943) from the University of Michigan and subsequently earned his two graduate degrees (M.S. '50, Ph.D. '57) in Bacteriology from the same institution.
He is a member of numerous professional organizations including: American Society for Microbiology, Pathologic Society of Great Britain and Ireland, Royal Society of Tropical Medicine and Hygiene, Association of Clinical Scientists, American Society of Clinical Pathology, Medical and Chirurgical Faculty of Maryland, Prince

PYRAMID POWER II - BIOGRAPHIES

George's County Medical Society, American Association of Bioanalysts, and certified by the American Board of Bioanalysts.

BORIS VERN, M.D.
Senior Medical Sciences Researcher

Dr. Vern is a physician with a Ph.D. in Biological Sciences, who is licensed to practice in the State of Maryland with a medical license pending in the District of Columbia. He is currently a Clinical Associate at the National Institute of Neurological Disease and Stroke, EEG Branch, National Institutes of Health, Commissioned Corps, U.S. Public Health Service, with the grade of Surgeon.

He did his pre-medical undergraduate work at Northwestern University (Evanston, Illinois) where he received a B.S. in 1968 and subsequently a Ph.D. (1971) in Biological Sciences. In 1972, he earned his M.D. degree from Northwestern University Medical School, Chicago, Illinois.

Dr. Vern, who is fluent in Russian and Ukranian, has the following publications to his credit:

"An Investigation of Possible Humoral Factors Related to the Initiation of Paradoxical Sleep in the Cat", Ph.D. Dissertation, *Dissertation Abstracts*, November 1971.

"Reinvestigation of the Effects of Gamma-Hydroxy-butyrate on the Sleep Cycle of the Unrestrained Intact Cat", *Electroencephalography and Clinical Neurophysiology*, 31 (1971), pp. 573-580. (Co-authored with John I. Hubbard.)

SKAIDRITE MALIKAS FALLAH
Research Psychologist and Linguist

Ms Fallah received her B.A. degree at Hunter College (New York - 1960), and her M.A. degree at the School for Advanced International Studies (SAIS), John Hopkins University, Washington D.C. in 1962.

She has received the New York Mayor's Scholastic Award, DAR Award, City Mildia County Scholarship, John Hopkins University Scholarship; one of 6 selected participants by the New York Board of Education for a two-month "Democracy Workshop" in Israel.

She has published over 17 research papers on various subjects.

CPSIA information can be obtained
at www.ICGtesting.com
Printed in the USA
LVOW07s0527290717
543061LV00001B/105/P